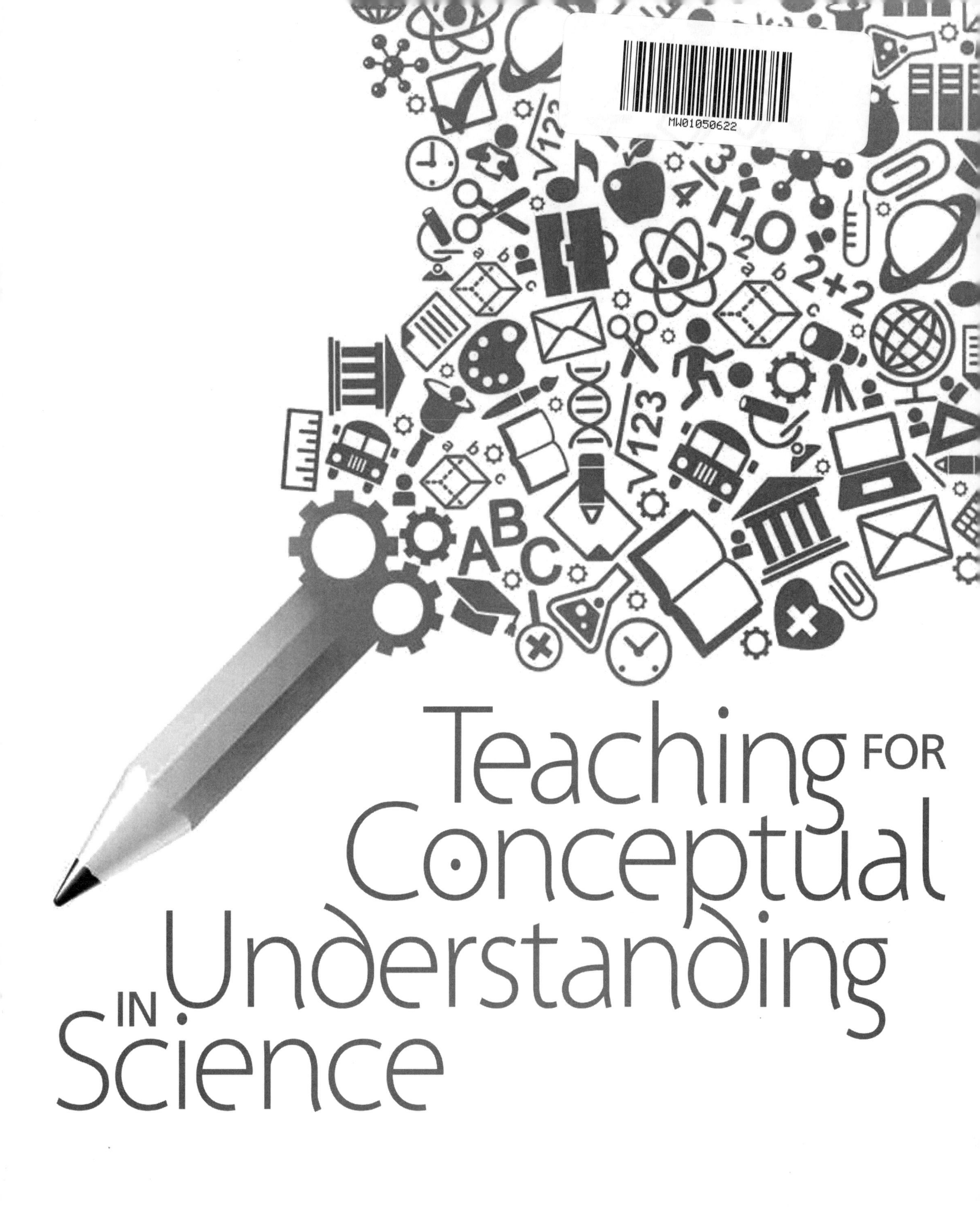
Teaching FOR
Conceptual
Understanding
IN Science

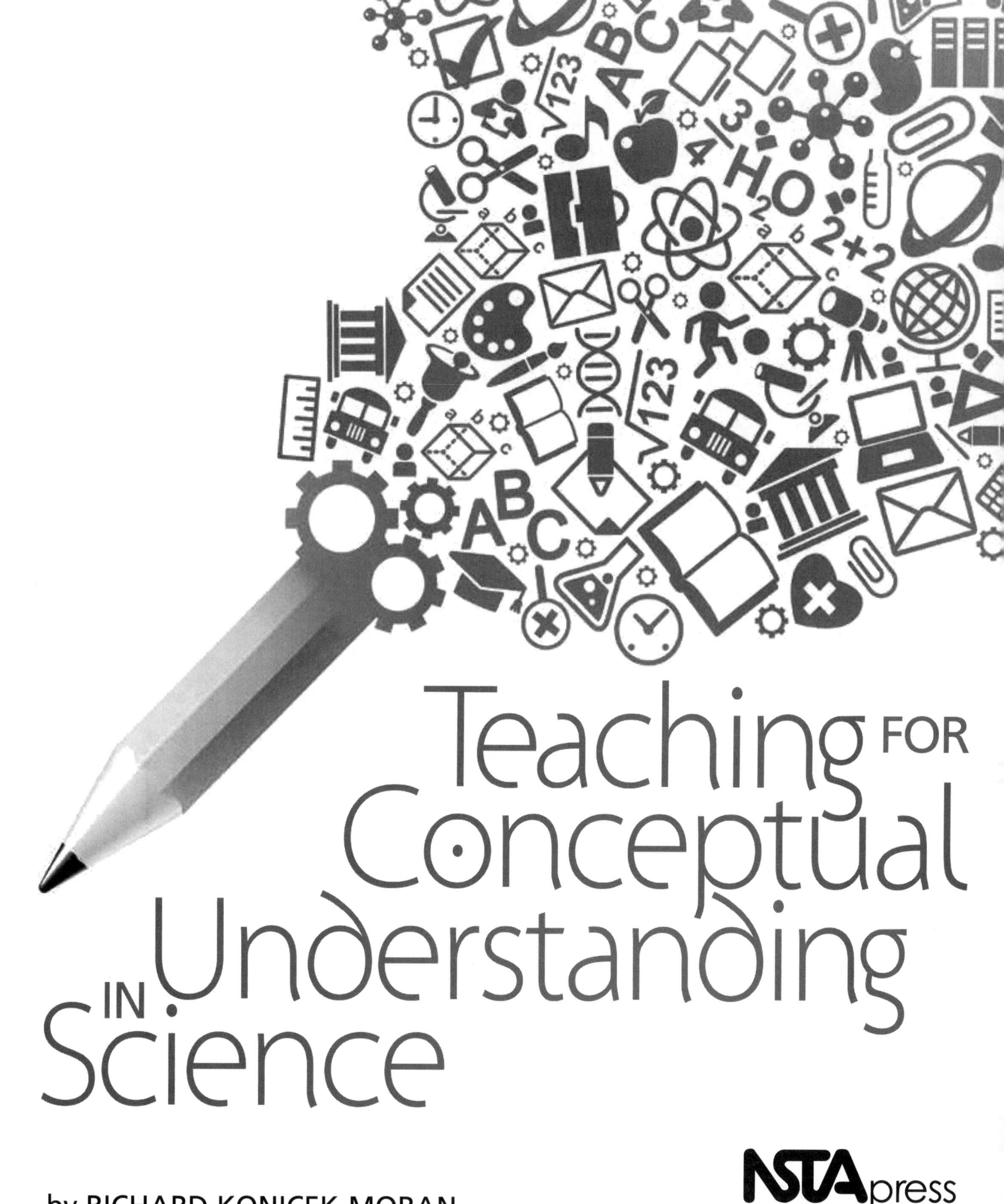

# Teaching for Conceptual Understanding in Science

by RICHARD KONICEK-MORAN
and PAGE KEELEY

Arlington, Virginia

Claire Reinburg, Director
Wendy Rubin, Managing Editor
Andrew Cooke, Senior Editor
Amanda O'Brien, Associate Editor
Donna Yudkin, Book Acquisitions Coordinator

Art and Design
Will Thomas Jr., Director

Printing and Production
Catherine Lorrain, Director

National Science Teachers Association
David L. Evans, Executive Director
David Beacom, Publisher

1840 Wilson Blvd., Arlington, VA 22201
*www.nsta.org/store*
For customer service inquiries, please call 800-277-5300.


19 18 17 16 5 4 3 2



**Library of Congress Cataloging-in-Publication Data**
Konicek-Moran, Richard.
Teaching for conceptual understanding in science / Richard Konicek-Moran and Page Keeley.
pages cm
Includes bibliographical references and index.
ISBN 978-1-938946-10-3
1. Science--Study and teaching. 2. Concept learning. 3. Concepts in children. I. Keeley, Page. II. Title.
Q181.K667 2015
507.1'2—dc23
2015001203

Cataloging-in-Publication Data for the e-book are also available from the Library of Congress.
e-LCCN: 2015001634

# Contents

# Preface

## How Did This Book Come to Be?

We have pondered over the topics covered in this book for years, presented workshops together, and talked for hours about how conceptual understanding can be achieved. Page had already acquired her passion for improving conceptual understanding using formative assessment tools. She has written many books as part of the *Uncovering Student Ideas in Science* series (Keeley, 2005–2013) and *Science Formative Assessment* (Keeley 2008, 2014). While at the University of Massachusetts in Amherst, Dick spent years studying and researching children's alternative conceptions in science, spending months working with Rosalind Driver, John Leach, Phil Scott, and other researchers at Leeds University in Great Britain, and then published his *Everyday Science Mysteries* series that contained, among other things, his collected thoughts about teaching for inquiry and conceptual understanding.

## Our Approach to This Book

Since we realized that our work had so much in common, we decided to try to put our thoughts and ideas gleaned from these experiences, research findings, and practices, into a book that would focus on this important topic. This book is a compilation of combined research findings, practices, and our personal experiences. It is woven into a conversational form of research notes, anecdotes, and vignettes showing how the principles of science might lead to better understandings. Although we have connected our writing to *A Framework for K–12 Science Education* (NRC 2012) and the *Next Generation Science Standards* (*NGSS*; NGSS Lead States 2013), this book is **not** meant to be a how-to manual for implementing the *NGSS* or other programs. There are and will continue to be ample numbers of publications written with those goals in mind. This book is, rather, a compendium, focusing on the major goal of science education for the 21st century and beyond—teaching for conceptual understanding. It is designed to be used with any set of national, state, or local science standards.

## Overview of the Book

There are 10 chapters in this book plus an appendix. In Chapter 1 we will address conceptual understanding: what it is and why it is important for teachers. Chapters 2 and 3 will focus on the history and nature of science and their importance to anyone teaching for conceptual understanding. Chapter 4 will present the current view of the nature of children's thinking, and Chapter 5 will look back at our attempts at making science teaching more meaningful through the use of the research findings available. Chapter 6 will examine *A Framework for K–12 Science Education's* learning practices of science and engineering (although our focus is primarily science) and their role in teaching science through the learning strands, while Chapter 7 is devoted to describing instructional models. Chapter 8 asks the question, what are some instructional strategies that support conceptual understanding? Chapter 9 will focus on connecting instruction, assessment, and learning. And finally, Chapter 10 will address learning in informal environments. In the Appendix is a case study of a lesson on balancing using the principles and ideas espoused in this book and in *A Framework for K–12 Science Education*. We include reflection questions at the end of each chapter for those readers who would like to extend their reading or thinking through book studies, professional learning groups, or science education courses as well as suggestions for resources available through NSTA that can be used to extend your learning.

## Audience and Uses for the Book

This book is written for practicing teachers, administrators, professional developers, and instructors of teachers, and future teachers themselves. This may seem like an all-inclusive broad audience, and that is intentional. Different parts of this book will appeal to different audiences. While you may read the book cover to cover, you may also choose to focus on specific chapters that best fit your purpose for using this book.

## Acknowledgments

We wish to thank the reviewers of our manuscript draft for their helpful suggestions and the staff of NSTA Press for their assistance and confidence in our ability to generate a book of this nature. To say that we are grateful to the teachers at all levels and the students of all ages who provided input is a gross understatement. This book is possible because of the extraordinary teachers and students we have had the privilege to work with. Of course, we acknowledge the patience and help of our spouses without whom we could not have produced this opus.

# About the Authors

**Dr. Richard (Dick) Konicek-Moran** is a professor emeritus from the University of Massachusetts in Amherst, a former middle school science teacher, and K–12 science district curriculum coordinator, involved in science education for the last 60 years. He is the author of the *Everyday Science Mysteries* series published by NSTA Press, and has been a life member of NSTA since 1968. Dick and his wife Kathleen, an internationally known botanical illustrator, have been volunteers with the Everglades National Park for the past 15 years and more recently for the Fairchild Tropical Botanical Garden in Coral Gables, Florida.

While studying to receive his doctorate at Teachers College, Columbia University in 1967, he authored or coauthored 25 textbooks in elementary and middle school science for the American Book Company. Since that time he has published dozens of articles in journals such as *The Kappan, Science and Children,* and others and given talks about his research at meetings of the American Educational Research Association (AERA). Many people have cited Dick's article "Teaching for Conceptual Change: Confronting Children's Experience," cowritten with Bruce Watson and published in the *Phi Delta Kappan* (1990), as seminal to the research and thinking of the movement to teach for conceptual understanding.

At the University of Massachusetts, Dick spent years researching the area of conceptual change while working with local and international school systems to foster improvements in science teaching. He spent just shy of a year in Dar Es Salaam in East Africa consulting with the Ministry of Education of Tanzania to improve science teaching at the college, elementary, and secondary levels. He also has led workshops in Quito, Ecuador, and in Dar Es Salaam, Tanzania, for teachers in the international schools. His invitation to visit the Peoples Republic of China in 1980 was one of the first educational visits to that country following the Cultural Revolution; he traveled, visited schools, and lectured at colleges in China. His time spent as a visiting research fellow at Leeds University in the UK with Rosalind Driver and the researchers at the Children's Learning In Science Group in 1990–1991 was arguably the most growth-inspiring time of his long career.

In 1978, he received the University Distinguished Teaching Award and the Distinguished Teaching Award from the Mortar Board Honors Society and in 2008 received the Presidential Citation from the National Science Teachers Association.

Dick enjoys music, playing jazz on keyboard, singing, and woodworking. He also enjoys nature and has volunteered at both Fairchild Garden (2 years) and the Everglades National Park (15 years). He is a full-time resident of Bradenton, Florida, in Manatee County.

**Page Keeley** recently "retired" from the Maine Mathematics and Science Alliance (MMSA), where she had been the Senior Science Program Director since 1996. Today she works as an independent consultant, speaker, and author, providing professional development to school districts and organizations in the areas of science, mathematics, STEM diagnostic and formative assessment, and conceptual change instruction. She has been the principal investigator and project director on three major National Science Foundation–funded projects, including the *Northern New England Co-mentoring Network, PRISMS: Phenomena and Representations for Instruction of Science in Middle School,* and *Curriculum Topic Study: A Systematic Approach to Utilizing National Standards and Cognitive Research.* In addition, she has designed and directed state Math and Science Partnership (MSP) projects including *Science Content, Conceptual Change, and Collaboration (SC4)* and *TIES K–12: Teachers Integrating Engineering Into Science K–12,* and two National Semi-Conductor Foundation grants, *Linking Science, Inquiry, and Language Literacy (L-SILL)* and *Linking Science, Engineering, and Language Literacy (L-SELL).* She developed and directed the Maine Governor's Academy for Science and Mathematics Education Leadership, which completed its fourth cohort group of Maine teacher STEM leaders, and is a replication of the National Academy for Science and Mathematics Education Leadership, of which she is a fellow.

Page is a prolific author of journal articles, book chapters, and 17 national bestselling books, including 10 books in the *Uncovering Student Ideas in Science* series, four books in the *Curriculum Topic Study* series, and three books in the *Science and Mathematics Formative Assessment: 75 Practical Strategies for Linking Assessment, Instruction, and Learning* series. She is a frequent invited speaker at regional and national conferences on the topic of formative assessment in science and teaching for conceptual change.

Prior to joining the Maine Mathematics and Science Alliance in 1996, Page taught middle and high school science for 15 years. She received the Presidential Award for Excellence in Secondary Science Teaching in 1992, the Milken National Distinguished Educator Award in 1993, and the AT&T Maine Governor's Fellow in 1994. Since leaving the classroom in 1996, her work in leadership and professional development has been nationally recognized. In 2008 she was elected the 63rd President of the National Science Teachers Association (NSTA). In 2009 she received the National Staff Development Council's (now Learning Forward) Susan Loucks-Horsley Award for Leadership in Science and Mathematics Professional Development. In 2013 she received the Outstanding Leadership in Science Education award from the National Science Education Leadership Association (NSELA). Page has led the People to People Citizen Ambassador Program's Science Education Delegations to South Africa (2009), China (2010), India (2012), and Cuba (2014).

Prior to teaching, Page was the research assistant for immunogeneticist Dr. Leonard Shultz at The Jackson Laboratory of Mammalian Genetics in Bar Harbor, Maine. She received her BS in life sciences from the University of New Hampshire and her master's in science sducation from the University of Maine.

You may note that in various places throughout the book, we will include some personal vignettes from our professional and personal lives that have relevance to the chapter. These will appear in a shaded box. We hope you find these anecdotes enjoyable and informative.

Thanks for coming along on this journey with us.

—Dick and Page

# Introduction

It has been more than four decades since David Ausubel made his famous and oft-quoted statement that the most important single factor in learning is *what the student knows*. He suggested that we find this out and teach the student *accordingly* (Ausubel 1963). And so, we have been trying for over four decades to find out what "accordingly" means. We have certainly made great progress in finding out what students bring to the classroom. Some will agree that it was the film *A Private Universe* (Schneps and Sadler 1987) that began the ACM (Alternative Concept Movement) of the 1980s and 1990s, resulting in a deluge of research about student alternative conceptions in science and mathematics. It also resulted in such books as the *Uncovering Student Ideas in Science* series and other attempts at familiarizing teachers with diagnostic and formative assessment, which helps us to focus on the ideas students bring to our classrooms and make informed instructional decisions. So we are now fairly competent in having the tools for finding out what our students know, but educators have found many different ways to bring about what has been commonly called "conceptual change."

This is what this book is about. How do we move our students from their present, limited knowledge of certain scientific concepts toward an understanding closer to what scientists now believe and that local, state, and national standards expect? Secondly, what does current research tell us about moving students toward deeper understanding of both science as a process and a set of practices and science as a knowledge base?

As Jean Piaget said in *Genetic Epistemology*, "Knowledge … is a system of transformations that become progressively adequate" (1968). By this we believe that he meant that knowledge in the broadest social community as well as in the personal realm is built over time and is subject to change until it becomes "adequate," at least for the time being. We have to realize that new theories are constantly being developed that help us interpret new data that is being collected every moment.

Philosophers have argued about knowledge for centuries and, as philosophers are wont to do, continue to argue about this concept *ad infinitum*. After all, that's their job. That's what philosophers do. Epistemology (a form of philosophy) asks three basic questions:

- What is knowledge?
- How do we come to know?
- How do we know what we know?

# Introduction

Sound familiar? We teachers ask these and other questions of our students and ourselves day in and day out. What are we teaching? Are the standards the last word in scientific knowledge? Where did the core ideas in the standards come from? How do we know what our students and we really know? Does it matter to us, as teachers, what students think? Where does the "knowledge" that they bring to our classrooms come from? Are their beliefs useful to us in our attempts to bring them closer to that which science deems the best explanations at this time? Do these ideas have value for future learning? Is it our role to try to change their ideas to match those of the standards to which students and teachers are held accountable? Is it really possible to try to switch their ideas for new ideas?

But we are getting ahead of ourselves. These questions dominate science education today, and we want to discuss them with you and acquaint you with some ideas that might help us to answer them.

Many philosophers and educators believe that we actively develop our own ways of trying to understand the universe in which we live. In other words, humans throughout history and well into the future, develop strategies to interpret the world so that it makes sense. We do not *discover* "truths" about our world; we develop explanations, test them for their ability to help us predict the behavior of the universe, and change them according to their ability to serve us. These ideas change over time, and each time they change, we come closer and closer to ideas that are more "adequate" than the last ones. Scientists in every field of endeavor are in a process of evolving ideas so they become more and more powerful in helping us to understand how our world works. We do this individually and socially. We do it as educators. We do it in science. We do it in economics. We do this in all of the major areas of thought and study. We put new theories out for public scrutiny and ask the societies of scientists, economists, historians, and others to evaluate them and see if they are acceptable and better in explaining our world than the older theories. Over time, if the new theories prove more powerful and useful, they replace the old ideas and thus, the disciplines evolve and knowledge grows.

Do we ever find "truth"? Perhaps the best answer is "yes and no." At this time, the ideas of an Earth that is a globe has been verified by modern technology that has given us a view from space that prior scientists who believed this were not privileged to have. However, we know that even though humans have walked upon the Moon, the data gathered there are still under scrutiny, and theories about the Moon, its origin, and its relation to the Earth are still being debated among societies of scientists called astronomers and geologists.

We posit that the same process goes on in life and in particular in schools, where children come to us with their own conceptions about what makes the world work. They do not appear before us with blank minds but with minds full of ideas that they have developed over their growing years that help them to understand, in their own way, what makes the world tick. Up to that point, their ideas were sufficient and allowed them to function, but then in school they are introduced to ideas that may be different from those they held before.

And thus, the problem is generated. These prior concepts are usually sound enough for the children to be comfortable with them; but we know that broader ideas—often those that seem, to the ordinary person, to fly in the face of all they know (what science educators call *counterintuitive*)—are more useful and powerful. Students' ideas are also thoroughly ingrained and persistent. How in children, just as in society in general, do these ideas change and become more useful? Let's look at the research, the history of science, the thinking, and the dreams that are leading us toward a better way to help children learn science and be active participants in science along with us who teach it.

# Chapter 1

## Teaching Science for Conceptual Understanding: An Overview

### What Do We Mean by Teaching for Conceptual Understanding?

A primary goal of science education is teaching for conceptual understanding. But what does this mean in an environment where scores on standardized tests are equated with student achievement in learning science? Do passing scores on standardized tests indicate students deeply understand science? Does filling students' heads with "mile-wide, inch-deep" information so they will be prepared for testing support conceptual understanding? Even when not faced with the pressures of testing, do our instructional routines get in the way of teaching for conceptual understanding? We argue that teaching for conceptual understanding can and should exist alongside the pressures of testing, "covering the curriculum," and instructional routines, if we change our beliefs about teaching and learning. But first we need to examine what conceptual understanding means.

Conceptual understanding is very much like making a cake from scratch without a recipe versus making a cake from a packaged mix. With the packaged mix, one does not have to think about the types and combination of ingredients or the steps involved. You make and bake the cake by following the directions on the box without really understanding what goes into making a cake. However, in making the cake from scratch, one must understand the types of ingredients that go into a cake and cause-and-effect relationships among them. For example, someone who understands baking knows that baking soda and baking powder are essential ingredients, understands the effect each has on the cake, how much to add of each, and when and how they should be added to the mixture in order to ensure batter uniformity. In other words, making the cake from scratch involves conceptual understanding rather than simply following a recipe.

Let's begin with the term *understanding*. One of the impediments to teaching for understanding lies in the way science instruction is sometimes delivered through direct instruction involving the passing on of information from the teacher to the student through techniques such as lecture, which involve little or no student interaction with the content. There is the story of the teacher who, upon seeing that most of the students had failed a test given at the end of a unit, responded, "I taught it, they just didn't learn it." The difference here, of course, is in the distinction between teaching and learning. Teaching does not automatically produce understanding. An important aspect of teaching is communication, yet "teaching as telling," even when combined with diagrams, computer simulations, and demonstrations, ignores how the student is making sense of the information if instruction is primarily focused on presenting information. A teacher can utter words and sentences, write symbols and equations on the board, use PowerPoint slides, and perform virtual or live demonstrations without effectively communicating ideas or concepts. In 1968, Robert Mager wrote, "If telling were teaching, we'd all be so smart we could hardly stand it" (p. 7).

Reading science textbooks, defining vocabulary, filling out worksheets, and answering low-level questions at the end of the chapter are also forms of passive instruction. These activities often involve pulling information from text with minimal intellectual engagement. The student may be able to reproduce the words or symbols she receives without understanding the meaning behind them or the power of using them to argue or predict and delve deeper into the ideas involved. People who are very good at memorizing facts and definitions often engage in what may be called *literal understanding.* Do you recall students who did well in school because they had eidetic or photographic memories? They could tell you what was on any page in the textbook or reproduce any graph or picture at a moment's notice exactly as it appeared in the book. Usually, because of the nature of testing, they scored very well. Yet, these students might not have been able to understand basic concepts that provide explanatory evidence for ideas about phenomena.

**Figure 1.1. "Wet Jeans" Probe**

**Wet Jeans**

Sam washed his favorite pair of jeans. He hung the wet jeans on a clothesline outside. An hour later the jeans were dry.

Circle the answer that best describes what happened to the water that was in the wet jeans *an hour later.*

**A** It soaked into the ground.

**B** It disappeared and no longer exists.

**C** It is in the air in an invisible form.

**D** It moved up to the clouds.

**E** It chemically changed into a new substance.

**F** It went up to the Sun.

**G** It broke down into atoms of hydrogen and oxygen.

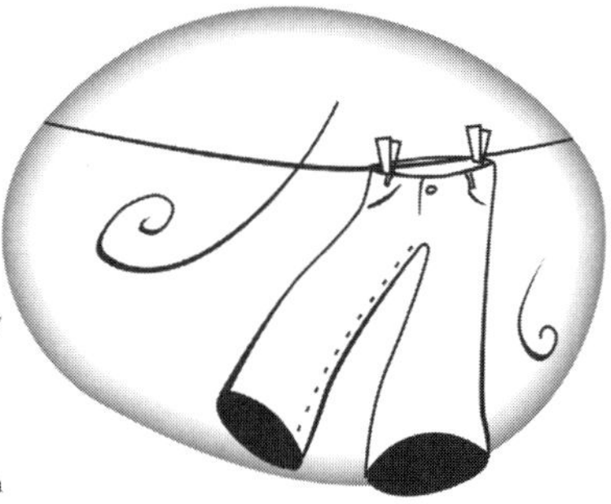

Describe your thinking. Provide an explanation for your answer.

*Source:* Keeley, Eberle, and Farrin 2005.

Take the concept of *evaporation* as an example. A student who is taught the water cycle may be able to recite word for word the definitions of *evaporation, condensation*, and *precipitation*. Furthermore, the student may be able to reproduce in detail a drawing of the water cycle, including a long arrow that points from a body of water to a cloud, labeled *evaporation*. On a standardized test, the student can answer a multiple-choice item correctly by matching the water cycle processes with the correct arrow on a diagram. All of this knowledge retrieved from memory may pass for understanding. However, when presented with an everyday phenomenon, such as the one in Figure 1.1, many students do not understand conceptually that when water evaporates, it goes into the air around us in a form we cannot see called water vapor (Keeley, Eberle, and Farrin 2005). They rely on their memorization of the term *evaporation* and the details of a water cycle diagram showing long arrows labeled *evaporation* to select distracter D: "It moved up to the clouds." The student lacks the conceptual understanding of what happens after water evaporates. This student may also have difficulty explaining why there is dew on the grass in the morning or why water forms on the outside of a cold drink on a hot summer day. The student may use the words *evaporation* and *condensation*, yet not understand where the water went or where it came from to explain a familiar phenomenon.

A typical routine in science classrooms is to assign a reading from a textbook or other source and have students answer a set of questions based on the reading. The text becomes the "deliverer" of information. Take for example, the passage, The Chemovation of Marfolamine in Figure 1.2.

**Figure 1.2. The Chemovation of Marfolamine**

> Marfolamine is a gadabolic cupertance essential for our jamination. Marfolamine was discovered in a zackadago. It was chemovated from the zackadago by ligitizing the pogites and then bollyswaggering it. Marfolamine will eventually micronate our gladivones so that we can homitote our tonsipows more demicly.

Now answer the following questions based on the passage:

1. What is marfolamine?
2. Where was marfolamine discovered?
3. How is marfolamine chemovated?
4. Why is marfolamine important to us?

Were you able to answer all four of the questions correctly, including the essential question in #4? Then you must know a lot about marfolamine! But do you understand anything about marfolamine? No, all you did was look for word clues in the text and parrot back the information. You did not need to intellectually interact with any of the concepts or ideas in the text. You did not share any of your own thinking about marfolamine. Probably you didn't need to think at all! While this is an exaggeration of a familiar instructional scenario, it is also typical of what some students do when asked to answer questions based on reading science text, especially text that is heavily laden with scientific terminology.

Lectures and recalling information from text are not the only instructional routines that fail to develop conceptual understanding. Picture the teacher who does a demonstration to show how the Moon's orbit around the Earth is synchronous with its rotation. The teacher provides the information about the Moon's orbit and rotation and then demonstrates it in front of the class using a lamp to represent the Sun, a tennis ball to represent Earth, and a Ping-Pong ball to represent the Moon. The students watch as the teacher demonstrates and explains the motion. But what if, instead, the teacher starts by asking students an

### Figure 1.3. "How Long Is a Day on the Moon?" Probe

**How Long Is a Day on the Moon?**

Four students were designing a Moon base for a science project. Planning the Moon base was easy. But deciding what a day-night cycle on the Moon base would be like was hard! All four students had different ideas. Here is what they said:

**Hannah:** "I think the length of the day-night cycle on the Moon is 24 hours."

**Sachet:** "It depends where the Moon base is. If it is on the dark side of the Moon, there will never be daytime."

**Ravi:** "I think there would be about two weeks of sunlight and two weeks of darkness."

**Manuel:** "It depends on the Moon phase. In a crescent Moon, daylight would be much shorter. When there's a full Moon, daylight would be much longer."

Which student do you think has the best idea? ________________ Explain why you agree.

*Source:* Keeley and Sneider 2012.

interesting question such as the one in Figure 1.3, listens carefully to their ideas, and then plans instruction that involves the students creating and using a model to figure out the best answer to the question? Clearly this example, which gives students an opportunity to think through different ideas and interact with a model used to explain the phenomenon, is more likely to result in conceptual understanding.

Teaching for conceptual understanding is a complex endeavor that science teachers have strived for throughout their careers. David Perkins, a well-known cognitive scientist at Harvard University, has been examining teaching for understanding for decades. He says that while teaching for understanding is not terribly hard, it is not terribly easy, either. He describes teaching for understanding as an intricate classroom choreography that involves six priorities for teachers who wish to teach for conceptual understanding (Perkins 1993):

1. Make learning a long-term, thinking-centered process.
2. Provide for rich, ongoing assessment.
3. Support learning with powerful representations.
4. Pay heed to developmental factors.
5. Induct students into the discipline.
6. Teach for transfer.

These teaching priorities identified two decades ago apply to current science teaching. In addition, recent research on learning in science is helping us understand even more what it means to teach for conceptual understanding in science. We will dive into past and present research and efforts to support teaching for conceptual understanding in science throughout this book, but first we need to define what we mean by a concept and explore factors that affect how we teach and learn science concepts.

## What Do We Mean by Concept?

The word *concept* has as many different meanings to science educators as the word *inquiry*. In this book, we equate it with a general idea that has been accepted by a given community. A. L. Pines defines a concept as "packages of meaning [that] capture regularities [similarities and differences], patterns or relationships among objects [and] events" (1985, p. 108). Joseph Novak, known for his research on concept mapping, similarly defines a concept as a perceived regularity in events or objects, or records of events or objects, designated by a label. The label for most concepts is a word, but it could be a symbol, such as % (Novak and Cañas 2006).

To give an example, *table* is a concept. Once a person has the concept of *table*, any object that fits a general description or has common attributes can be called a table. It may have three legs, be round or square or rectangular, or sit on the floor as in a Japanese restaurant. It may be made of many substances. But if we have internalized the concept of *table*, we know one when we see it. The same would be true of the concept *dog*. Whether it is a St. Bernard or a Chihuahua, we know a dog when we see one. Before a child is familiar with the superordinate concept of *dog*, she may call any furry four-legged animal a dog. But once she has internalized the characteristics of "doggyness" she recognizes one, regardless of breed.

A concept is an abstraction. Tables did not come into this world labeled as such. In fact, depending on where you live in this world, a table is called by many names, depending on which language you use. However, whatever the language, whatever the name, the concept of *table* remains the same in all cultures. The concepts of table or dog are *constructions* of the human mind. A concept is basically a tool constructed for the purpose of organizing observations and used for the prediction of actions and classification.

In science, we use fundamental building blocks of thought that have depth and call them *concepts*. Words, such as *energy*, *force*, *evaporation*, *respiration*, *heat*, *erosion*, and *acceleration*, are labels for concepts. They are abstractions developed in the minds of people who tried to understand what was happening in their world. Concepts may also consist of more than one word or a short phrase such as *conservation of energy*, *balanced and unbalanced forces*, *food chain*, or *closed system*. Concepts imply meaning behind natural phenomena such as phases of the Moon, transfer of energy, condensation, or cell division. When we use a concept, there is usually some understanding of what is associated with it. For example, *condensation* is the concept. It conjures up an image of water drops formed on an object. The concept becomes an idea when we try to explain or define it. For example, the concept of *condensation* becomes an idea when we associate water vapor in the air reappearing as a liquid when it comes in contact with a cool object. It becomes a definition when we define condensation as the conversion of water in its gaseous form to a liquid. Concepts are the building blocks of ideas and definitions. Another way to distinguish concepts from other ways to express one's thinking is to imagine that a teacher asks a student what is in her backpack. The student replies, "my school books, some supplies, and snacks." These are concepts that imply meaning of the kinds of things the student has in her backpack rather than saying, "my biology textbook, my social studies book, my math book, two notebooks, pencils, pens, assignment pad, a granola bar, a bag of chips, and an apple." Behind all concepts in science are data, a history of observation and testing, and a general agreement of scientists within any given domain.

When students have an *understanding* of a concept, they can (a) think with it, (b) use it in areas other than that in which they learned it, (c) state it in their own words, (d) find a metaphor or an analogy for it, or (e) build a mental or physical model of it. In other words, the students have made the concept their own. This is what we call *conceptual understanding*.

## Learning to Speak and Understand a New Language

Words and symbols are important. Language is the way of communicating science concepts, but the language of science is not always the language of everyday life. Language can affect how we think about concepts in science. Often, a word or symbol has a special meaning to a scientist, different from the way a nonscientist may use the word. A scientist knows what is meant when someone says, "Close the door—you're letting the cold in, " even though she or he understands that in thermodynamics, that there is no such thing as "cold" and that heat always moves from warmer to colder areas. The scientist has conceptual understanding that overrides the incorrect terminology. The same is true of "sunrise" or "sunset" which is really the *illusion* of the apparent motion of the Sun in the sky. Someone with a conceptual understanding of the phenomena understands that it is the Earth's rotation that is responsible for this visual effect. Some concepts used in the science classroom are counterintuitive to students' ideas. For example, the definition in physics of *acceleration* can mean slowing down as well as speeding up (or changing direction). This does not make sense to students based on their everyday encounters with the word *acceleration*, which to them means going faster. After all, don't you make the car go faster by pushing down on the "accelerator"?

Many of us live or work in areas with increasingly diverse populations. For example, the authors of this book both live in Florida for part of the year. This often means that people who speak a language different from our first preference surround us. If the trend continues, a majority of the residents that make up our neighborhoods may speak Spanish. To communicate effectively, we may need to learn Spanish and to become bilingual. It takes perseverance and a desire to think in a new language, rather than merely translate word for word. Instead we must learn dialog, cadence, colloquialisms, a new vocabulary, and most importantly, culture. Phrases cannot be taken literally when translated from one language to another. For example, someone might say "So long" meaning "goodbye," which makes no sense, if you think about it literally. Speaking science is very much the same but can pose even more problems.

Speaking science has an added difficulty for students. One problem is that in colloquial language a scientific word may have a different meaning altogether, which affects our understanding of the concept underlying the word. For example, you might hear someone say, "Oh, that's just a theory," meaning that it is just a guess or unproven idea, when in science *theory* means a well-supported explanation of phenomena, widely accepted by the scientific community. People who recognize both the scientific concept of a theory and the way the word is used in our everyday language can accommodate the two meanings, but this is not the case with students new to the language of science. As science teachers, we need to be aware of the differences in meanings between our students' daily use of certain words and the scientific meaning of these same words. Another problem is that the language of science is tied directly into the practices and rules of science and therefore is tied to experience within the discipline. Students need to experience the practices of science in order to understand conceptually the language that is used.

Many teachers use the technique of "word walls." This technique is often used in classrooms with English as a Second Language (ESL) students, but it is an effective way to introduce vocabulary in context to all students—and if arranged in an interactive way, it is also a way to organize concepts into instructional plans so science is not treated as vocabulary but rather vocabulary is introduced for the purpose of communicating scientific ideas.

Traditional word walls have objects or pictures of objects and their names posted on the wall to help students become familiar with new words that represent a concept. The interactive word wall is an organic, growing wall that is planned by the teacher but developed with the help of the students. The class adds ideas and objects to the wall with the help of the teacher, and as the unit grows toward completion, the wall grows to include the newest concepts and the objects and ideas that go with it. The word wall is used to develop understanding, as students organize words for deeper conceptual meaning. Conceptual teaching strategies such as word walls will be explored further in Chapter 8.

Although vocabulary is important for science learners, we must remember that words are not science. Zoologists do not study words but use words to communicate their study of animals with others who share the same vocabulary. Vocabulary needs to be introduced to students in the midst of their engagement with objects. Since we advocate hands-on, minds-on science activities, the time to introduce vocabulary is either during the activity or during the discussion afterward. For example, when learning about the motion of pendulums, the word *amplitude* would be introduced while the students are investigating whether the pendulum's motion is changed when the pendulum is allowed to travel through a smaller or larger arc.

Science as a discipline has words and symbols that have specific meanings. Think of scientific fields that deal with symbolic structures like genome sequencing. Math, too, uses symbols to express ideas and concepts. Understanding the nature of science presents challenges in the way we use language and symbols. Let's take a look at some of the most important words and phrases that often have a popular *double entendre* when used to describe the nature of science. Please note that the descriptions provided below are a simplified view of the nature of science. Philosophers and linguists might argue about each of these points, but for the purpose of helping you, the teacher, understand the language of the nature of science in the context of K–12 education, we hope these points and descriptions will suffice.

### Theory

As we mentioned above, in everyday speak, this word may mean a hunch, an opinion, or a guess. In science, it means an idea that has been tested over time, found to be consistent with data, and is an exemplar of stability and usefulness in making predictions. A theory explains why phenomena happen. You may hear people say, "I have a theory that the Chicago Cubs will win the World Series next year." This is usually based on a belief system grounded in a preference steeped in loyalty (and sometimes fruitless hope). Unfortunately for Cubs fans, there are few data that will support this "theory." You may also hear someone dismiss the theory of biological evolution as "just a theory." You can be assured that this person has a lack of conceptual understanding about how a theory in science is tested, rooted in evidence, and thus held in the utmost respect by the scientific community as being accurate and useful. We see evidence of biological evolution happening every day. Bacteria evolve into drug-resistant strains and animals and plants adapt to changing environmental conditions over time. The theory of natural selection attempts to explain how this happens, and it does this quite successfully. Figure 1.4 is an example of a formative assessment probe used to elicit students' (and teachers') conceptual understanding of a scientific theory. The best answers are *A, D, G,* and *I.* The distracters (incorrect answer choices) reveal common misunderstandings people have about the word *theory* as it applies to science.

### Hypothesis

A hypothesis in science is often an "if ... then" statement in response to a scientific question that provides a tentative explanation that leads an investigation and can be used to provide more information to either strengthen a theory or develop a new one. A hypothesis is a strongly developed prediction, based on prior observation or scientific knowledge that if something is done, an expected result will occur. It is constructed with a great deal of planning and a reliance on past evidence. Some educators use the term *educated guess* to describe a hypothesis. This is another example of the misuse of language. There is no guesswork involved in developing hypotheses and using language in that way incorrectly portrays the concept of a hypothesis.

In science, a hypothesis is never a "sure thing" and scientists do not "prove" hypotheses. Students who complete an investigation and claim that their results prove their hypothesis should be encouraged to say their results support their hypothesis. Scientists learn from hypotheses that are shown to be wrong as well as those that provide expected results. Science teachers are often guilty of asking children to hypothesize something that cannot be more than just a wild guess or unsubstantiated prediction. Students should learn that a hypothesis should be acceptable only if there is preliminary evidence through prior observation or background knowledge to back up the hypothesis.

**Figure 1.4. "Is It a Theory?" Probe**

# Is It a Theory?

Put an X next to the statements you think best apply to scientific theories.

_____ **A** Theories include observations.

_____ **B** Theories are "hunches" scientists have.

_____ **C** Theories can include personal beliefs or opinions.

_____ **D** Theories have been tested many times.

_____ **E** Theories are incomplete, temporary ideas.

_____ **F** A theory never changes.

_____ **G** Theories are inferred explanations, strongly supported by evidence.

_____ **H** A scientific law has been proven and a theory has not.

_____ **I** Theories are used to make predictions.

_____ **J** Laws are more important to science than theories.

Examine the statements you checked off. Describe what a theory in science means to you.

*Source:* Keeley, Eberle, and Dorsey 2008.

## Author Vignette

I recently worked with a group of middle school teachers, using the formative assessment probe "What Is a Hypothesis?" to uncover their ideas about the word *hypothesis* (Keeley, Eberle, and Dorsey 2008). Using the card sort technique, the answer choices were printed on a set of cards and teachers sorted them into statements that describe a scientific hypothesis and statements that do not describe a scientific hypothesis.

### What Is a Hypothesis?

Hypotheses are used widely in science. Put an X next to the statements that describe a hypothesis.



_____ **A** A tentative explanation

_____ **B** A statement that can be tested

_____ **C** An educated guess

_____ **D** An investigative question

_____ **E** A prediction about the outcome of an investigation

_____ **F** A question asked at the beginning of an investigation

_____ **G** A statement that may lead to a prediction

_____ **H** Included as a part of all scientific investigations

_____ **I** Used to prove whether something is true

_____ **J** Eventually becomes a theory, then a law

_____ **K** May guide an investigation

_____ **L** Used to decide what data to pay attention to and seek

_____ **M** Developed from imagination and creativity

_____ **N** Must be in the form of "if…then…"

Describe what a hypothesis is in science. Include your own definition of the word *hypothesis* and explain how you learned what it is.

The best answer choices are *A, B, G, K, L,* and *M*. Almost all of the teachers selected *C* and *I* as statements that describe a scientific hypothesis. As we debriefed and discussed, the teachers were adamant that *C* and *I* accurately described a scientific hypothesis. One teacher even took the group over to her classroom to point out *The Scientific Method* bulletin board she had in her classroom made up of purchased placards that depicted stages of the scientific method, including the one shown below that implies a hypothesis is an educated guess:

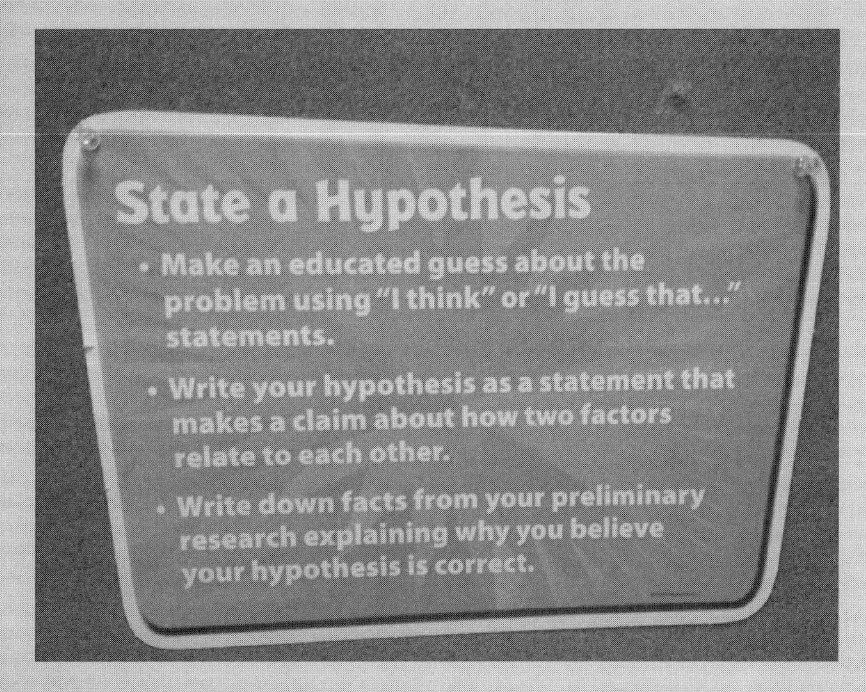


Furthermore, I noticed another placard titled, "Analyze/Make a Conclusion," in which the last bulleted suggestion was, "If the results prove your hypothesis to be correct, perform the experiment again to see if you get the same results." No wonder some teachers hold these misunderstandings! We discussed the need to be aware of these misrepresentations of the nature of science when purchasing and displaying materials such as these that further perpetuate students' misuse of words such as *prove* (a better choice is *support*) or *educated guess* when referring to hypotheses.

—Page Keeley

### *Data*

*Data* is the plural form of *datum*. Data are a collection of observations or measurements taken from the natural world by means of experiments or the observation of information that shows a consistent pattern. One of the most consistent errors (even in the public media) is to fail to differentiate between the singular and the plural forms of this word. Data *are* and a datum *is*. Data do not come with an inherent structure. According to *Ready, Set, SCIENCE!* structure must be imposed on data. By this the authors mean that data can be processed in many ways but they must be organized and reorganized to answer questions. Using data correctly is one of the most important lessons students can learn in science (NRC 2007).

### *Evidence*

This term is used to describe a body of data or a base that shows consistent correlations or patterns that become the basis for a *scientific claim*. Observations and experience lead to claims. A *claim* in our everyday language can be an opinion or belief. *Scientific claims* are always based on scientific evidence and educators should make this clear to students who are making claims. In other words, a reasonable response to a child who makes a claim would be, "What is the scientific evidence that supports your claim?"

For example, I notice in the morning that my car is covered in water droplets. I could make a claim that it has rained, but it really is not a scientific claim until I have searched for other evidence. Has anyone watered the area with a hose overnight? Has relative humidity had anything to do with the water droplets? Is there water on anything else but the car? Could the water come from dew? I must take into consideration many more factors before I can make a scientific claim. Whenever a student makes a claim in a classroom, the teacher must ask for evidence supporting it. After time, claims made will become more carefully considered, and claims backed by scientific evidence will become common practice.

### *Experiment*

There is a tendency for people to refer to any activity involving science that occurs in a classroom as an experiment. This is an overgeneralization. All experiments are investigations, but not all investigations are experiments. Experimentation is a process in which variables are identified and conditions are carefully controlled in order to test hypotheses. Think of all of the things that must be done before an experiment can be designed and carried out: Students first develop a true hypothesis that is based on sufficient evidence and claims. The experimental hypothesis will most likely have an "if ... then" statement and will be set up with all available variables controlled so that the data collected can lead to a definitive answer. For example: "*If* I change the length of the pendulum *then* the period of the pendulum will change." To test this idea, one must keep the mass and shape of the bob the same, and use the same angle of release. The only thing changed is the length of the string.

Featured one morning on the Weather Channel was a physical model made at the Massachusetts Institute of Technology (MIT) that showed how the air currents of different temperatures were affected by the rotation of the Earth. Unfortunately, in their exuberance about how the model explained what they were showing on their maps, the hosts of the show called the demonstration an *experiment*. Here we are again being treated to the kind of everyday—but for our purposes—sloppy language usage that permeates our society and helps to confuse the meaning of science concepts. One of our prime targets in correcting the language of science should probably be the national media.

## Learning the Language of Science Education

Even the terminology we use as science educators to describe conceptual understanding may be unfamiliar language to some teachers. The following are a few of the important words used to describe conceptual teaching and learning that we will use throughout this book:

### *Alternative Conceptions*

Basically, *alternative conceptions* are mental models conceived by individuals to try to explain natural phenomena: "The Moon phases are caused by shadows." "Density is caused by how tightly packed the molecules in matter are." "When water appears on the outside of a glass in warm, humid weather, the glass is leaking. "Cold creeps into a house if there are leaks in the structure." "Metal objects are always cooler than wooden objects, even when they are in the same room for a long time." These are all examples of alternative conceptions or, as some would call them, *misconceptions*. They are incomplete theories that people have developed to try to understand their world. By "incomplete," we mean that they are not fully thought out and have limited use. Misstating the number of chromosomes in the human cell (which happened in textbooks in the 1950s) is not an alternative conception; it is merely misinformation. For a statement to be an alternative conception, it must be a theory that is used to explain a phenomenon, and is usually self-discovered by a person trying to explain that phenomenon.

Example: A person who has heard the term *population density* will probably first apply the idea of tightly packed individuals to scientific ideas of density. If she does not realize that atoms have different masses and that packing does not cause the difference in mass in objects of the same size, she will have a completely erroneous conception of molecular mass and density. Holding on to this alternative conception will make it very difficult to think of *density* in the accepted scientific paradigm. Children (and adults) are perfectly capable of holding on to several theories at the same time without seeing them as contradictory. As science teachers, we have an obligation to try to see the world through a child's eyes, to listen to their conceptions and use them to introduce the child to other ways of viewing the world.

### *Conceptual Change*

Throughout history, ideas have been debated, and every so often old ideas are either put aside or modified in order to match current observations, data, or the need to explain phenomena in a more useful and simpler way. As we will discover in Chapter 2, sometimes change has happened smoothly and other times, a revolution in thinking has occurred (Kuhn 1996). In many cases, the older ideas do not "go quietly into this good night" (apologies to poet Dylan Thomas!). Those of us who have used a theory or concept with success are loathe to giving it up to a new idea unless we are convinced that the newer idea is better in every way and explains phenomena more cogently. Einstein's theories of special and general relativity took years to be a dominant paradigm in physics.

In order to participate in conceptual change, we must be convinced that another explanation that uses the concept is more useful. The same is true of children who enter our classrooms with concepts they have used, possibly for years, with great success. Why would they want to change them unless they were seen to be no longer useful? Children's naive conceptions are built individually but are strongly affected by social and cultural conditions. They are not fully developed, but they work for the children and form a coherent framework for explaining the world.

For example, imagine a middle school child observing the Moon's phases changing each night. She cannot ignore the phenomenon, and therefore forms her own theory to explain it. The child has had previous experiences interacting with objects through play and other activities where she observes when an object blocks light from the Sun, a dark shadow is cast on the ground by the object. Part of the area around the object is in light, part is in shadow. The child uses this experience to develop a personal theory for the phases of the Moon by explaining that the Earth blocks part of the sunlight shining on the Moon and casts a shadow on that part of it.

Shapiro says it best in her book *What Children Bring to Light*. "When we teach science, we are asking learners to accept something more than scientifically verified ideas. We are asking them to accept initiation into a particular way of seeing and explaining the world and to step around their own meanings and personal understandings of phenomena into a world of publicly accepted ideas" (1994, xiii).

This is not always easy, as we know from experience. We will discuss this aspect of teaching further in subsequent chapters.

### *Paradigm*

In Thomas Kuhn's landmark book, *The Structure of Scientific Revolutions*, he says "[Paradigms are] examples of actual scientific practice—examples of which include law, theory, application and instrumentation together—provide models from which spring particular coherent traditions of scientific research" (1996, p. 10). Some examples of paradigms in science are: Ptolemaic astronomy, Copernican astronomy, and Newtonian corpuscular optics.

If you are a scientist, your research is influenced by and committed to a particular paradigm, and you follow certain rules and practices in your research dictated by that paradigm. For example, a dominant paradigm of Western science in the middle ages was that Earth was the center of the universe and that celestial bodies such as the Sun moved around the Earth. Can you imagine being a disciple of the new Copernican paradigm in, say, the 1540s that stated that the *Sun,* not the Earth, was the center of the universe, and deciding to do research in this "heretical" idea? Its influence would have probably made you work in secret for fear that the Roman Catholic Church of that time would excommunicate you or worse. Today, Copernicus's heliocentric theory is regarded by the Roman Catholic Church and scholars as one of the great revolutions in science.

Kuhn goes on to theorize that the history of science is rife with what he termed "paradigm shifts," during which time new paradigms influenced groups of converts and changed the whole nature of scientific thought (1996). In the same way, it may take a "revolution" in thinking to shift the paradigm that forms the basis for a person's alternative conception. (Carey 2009). We'll examine paradigms in depth in Chapter 2.

## Author Vignette

I remember when I was in graduate school, one of the required readings in our seminar class was Kuhn's *The Structure of Scientific Revolutions.* I initially found it to be rather wordy and challenging. I had to read a chapter several times, through sheer drudgery, in order to understand it. My first reaction was negative—why read such a dense, philosophical book if not to help me fall asleep with ease? Why can't we read something more modern and applicable to science teaching? How is this going to help me be a better science educator? After a couple chapters—and the first discussion we had in class, artfully facilitated by our professor—I became enthralled and enamored by this book. The term *paradigm*, which I had encountered in the popular lexicon, had new meaning for me, as did *revolution* and the term *normal science*. I was particularly interested in how Kuhn described the process of how one paradigm can replace another. Through our seminar discussions, my view

> of the nature of science was reshaped—I experienced my own paradigm shift as my assumptions about the scientific enterprise and words I used to describe it were challenged! Three years ago, I had to smile when my son gave me a copy of the book at Christmas. He had read it in one of his graduate courses and thought I would enjoy it (little did he know that I had to read it in one of my courses decades before). Today, this book sits on my shelf as one of the most important contributions to understanding the history and nature of science. As a science educator, I frequently see Kuhn's landmark book cited in the education literature on the nature of science. Perhaps it is one of the best and most authentic descriptions (albeit wordy and dense) of the nature of science that every science educator should read.
>
> —Page Keeley

### *Crosscutting Concepts*

One of the major concerns in learning any subject is that of organizing our thinking around major topics for easier retrieval and transfer of learning to the many related areas of a domain of knowledge. One of the secrets to internalizing knowledge is seeing its relationship to a larger, more encompassing set of ideas. Relationships among ideas give them credibility and help us all to group big ideas into larger, more comprehensive groups. If we can see that periodic motion can be used with the pattern of the planets and moons in our solar system and the motion of a pendulum or a reproductive cycle, we can see how they fit together. After all, science is all about finding patterns and using those patterns to explain the behavior of our natural world. *A Framework for Science Education* (NRC 2012) and the *Next Generation Science Standards* (NGSS Lead States 2013) identify the *crosscutting concepts* all students should master by the time they finish grade 12:

- Patterns
- Cause and effect: Mechanism and explanation
- Scale, proportion, and quantity
- Systems and system models
- Energy and matter: Flows, cycles, and conservation
- Structure and function
- Stability and change

If our students were to be familiar with these crosscutting concepts and be able to organize their learning in these groupings, transfer of knowledge and retrieval of information would become much more efficient.

### Models

The authors of *Ready, Set, SCIENCE!* define *models* as things that make our thinking visible (Michaels, Shouse, and Schweingruber 2008). When some people hear the word *model*, they think of a physical representation that is built to look like the real thing. But models are more than just physical replicas. For example, mental models are those we hold in our minds to try to explain the phenomena we see daily. They are personal models. For example, some young students have a mental model of the Earth, which allows them to understand why they seem to be on level ground although they may believe that the Earth is a sphere. Their model either has them in the center of the globe on a flat surface or standing on a flat part of the Earth within the round Earth. Early scientists like Ptolemy had a mental model that eventually became a conceptual model for his peers that specified Earth was the center of the planetary system. This conceptual model remained for many years because it corresponded to their observations that the Sun appeared to move across the sky and was consistent with the views of the Roman Catholic Church at that time. It took centuries before scientists such as Copernicus and Galileo had the courage to oppose the dominant model of that time and create their own mental models that showed that the Sun was the center of the planetary system.

Models can be mathematical, physical, conceptual, or computer generated. Models are often developed to try to approximate the real thing in a form that can be manipulated and studied in cases when a real situation cannot. Models also help students clarify and explain their ideas. The common classroom activity that involves building a replica of a cell out of food items or representing parts of an atom using cereal contributes to students' understanding of models as replicas made out of "stuff." While these may be representations that are not much different from 2-dimensional drawings, students seldom use them to explain their ideas or manipulate them to make predictions. In essence, they often fall more in the realm of arts-and-crafts projects than scientific models. Having a conceptual understanding of what a model is and is not is just as important as developing and using models in science.

We hope that looking at these examples of words we use to describe science and the understanding of science will be helpful to you as you think about designing instruction for conceptual understanding. We must realize that we are asking students to "step around" their own mental models and accept those ideas that are now considered the publicly accepted ideas (Shapiro 1994). They must also be aware that there may be "revolutions" in thinking and that paradigm shifts may occur in science during their lifetimes. This does not make science look weak, but helps us to see that scientific knowledge evolves. It is the nature of the discipline and its strongest attribute. Scientific knowledge is not dogma but a continuously changing set of ideas that are undergoing never-ending scrutiny by the members of the society we call scientists. We will explore this in more depth when we look at the nature of science in Chapter 3.

## From Words to Listening for Conceptual Understanding

One of the most important watchwords for teaching for conceptual understanding will be *listening.* A student's alternative conceptions are very important, and teachers need to be able to understand what the student is thinking. Alternative conceptions, no matter how naive or seemingly incorrect, are the foundations for building new and more complete conceptions. They provide us with a place to start teaching and with the information necessary to plan next steps.

Because of this, one of the most important best practices that has come to the forefront is *diagnostic and formative assessment* for the purpose of understanding student thinking and making decisions based on where students are conceptually in their understanding. One of the authors of this book (Page Keeley) specializes in science diagnostic and formative assessment. As a nation, we have been so extremely invested in summative testing since the advent of No Child Left Behind (NCLB) that some educators have often referred to it as No Child Left Untested. We agree that it is necessary and important to test for achievement and accountability, but it is evident that unless teachers know where their children are in their current conceptual development, they cannot plan for helping their students make changes in thinking as they design and facilitate instruction. This requires listening and responding to children when they think out loud. In order for us to hear them out loud, we have to give them a chance to tell us about their thinking and explain their ideas. We will address the topic of diagnostic and formative assessment and "science talk" in more detail when we get to Chapters 8 and 9 in this book.

One important researcher who addresses the issue of listening to children is Bonnie Shapiro from the University of Calgary. In her book *What Children Bring to Light,* she examines a fifth-grade classroom and the real responses of children to a vigorously taught series of lessons about how we see. In her research, she found that in her sample of six children, all but one did not believe what the teacher said, even though they successfully passed the unit by filling out their worksheets and completing their tests. The teacher never knew it because he didn't listen or probe the children's thinking. We'll examine Shapiro's research more fully in Chapter 4.

Often, when children and adults talk to each other, there is a problem of *incommensurability*. This term means, simply, that two people in a conversation are not speaking the same "language." Thomas Kuhn referred to this problem when he described a similar problem in the history of science (1996). Not only are teacher and student using different language, but also they are operating in different paradigms or rules about how the world is seen and studied. The students notice different things and focus on different questions than do adults. The teacher must be the one to try to overcome this incommensurability.

Two philosophers, Paul Thagard and Jing Zhu (2003), point out that there can be different emotional valences (i.e., weights or connotations) to incommensurability in conceptual understanding. They state that the concepts *baby* and *ice cream* have positive valences for most people, while the concepts *death* and *disease* have negative valences. People in the media, who are adept at "spin"—using language that makes their clients appear as positive as possible—have long been aware of this. Thagard and Zhu note that in order for conceptual change to occur, especially in emotionally charged areas of thought, each of the communicants would have to change their valence on the issues from negative to positive. They give as an example a Darwinian evolutionist and a creationist trying to reach a common ground. In order for each to achieve commensurability, each would have to change their emotional valence, and this may be very difficult, even impossible (look at our own ideologically charged political system). But it is important for teachers to be sensitive to the emotional impacts that the curriculum might be presenting to the children and be aware of the language they can use to change emotionally-based concepts to more evidence-based concepts.

Teachers often feel committed to changing a "wrong" idea as quickly as possible by whatever means they have at their disposal. Instead, since you are cast in the role of teacher-researchers we suggest that this is the time to listen as carefully as possible and to question the student(s) to find out as much as possible about where the ideas originated and how deeply the student(s) are committed to the idea to explain certain phenomena. Make them see how interested you are in how they think, and you will encourage them to consider their own thinking, and engage in what is known as *metacognition* (thinking about their thinking). The conversation does not have to be one-on-one. Instead, we suggest that students talk to each other and the teacher out loud, bringing students' thoughts to the front so all students can hear. Teachers have found that when they concentrate on the conceptual history of the group, the groups itself remains interested, even when the conversations may involve only a few members.

## Intentional Conceptual Change and a Community of Learners

This leads us to consider the recent pedagogical theory on *intentional conceptual change.* If we believe that both scientists and science learners gain knowledge in a community and that that knowledge is defined as a community consensus, it leads toward a belief that the teacher and the students are most effective when there is an intent to learn or change on the part of the learner and the community of students are goal-oriented toward understanding a new idea. When there is peer support and encouragement for learning, there is an atmosphere more conducive to conceptual change and understanding (Sinatra and Pintrich 2003). This may certainly lead us to building a community of learners as recommended by Bransford, Brown, and Cocking (2000). Hennesey suggests that metacognition (thinking about one's own thinking) is a primary ingredient in the working of a community of learners, stating that students have to be aware of how they came to their own knowledge claims before they can discuss them with others (2003). These knowledge claims raise the question of how to create the community of learners (including the teacher as learner) and conduct a class where the community is motivated toward solving a common question or problem. However, as Vasniadou points out, making an assumption that students will intentionally create strategies for developing intentional learning might be rather optimistic (2003).

We all know students can develop strategies for completing simple school-type tasks. It takes more effort to create the kind of atmosphere and curriculum that "grabs" the students and entices them into wanting to develop an inclusive community, intent on solving a common problem. One of the authors of this book (Dick Konicek-Moran) has published a series through NSTA Press called *Everyday Science Mysteries*. These mystery stories describe a common problem that can be used to motivate and capture the interest of all students in the class. The series provides open-ended stories that require metacognition and inquiry to find the best solution to the problem.

The following personal author vignette describes how a community of learners helped each other solve a common problem:

## Author Vignette

I once worked in a fifth-grade classroom in New England where the students had shown a great deal of interest in the apparent daily motion of the Sun. This came about through the reading of the story, "Where are the Acorns?" This story is about a squirrel that buries acorns using the Sun's effect on tree shadows during the Fall to predict where the acorns will be during the winter season (Konicek-Moran 2008).

Since the shadows change in the story, the students organized their own curriculum to find out as much as they could about the apparent movement of the Sun on a daily basis as well as seasonally. They predicted that the Sun would cast no shadow at midday (a common naive conception). They had already decided, through experimentation and discussion that midday was not necessarily noon but could be defined as a point halfway between sunrise and sunset. The children needed to find the midday point for a given day. They chose to use the tables in *The Old Farmer's Almanac*. Mathematically, this is not as easy a task as it might appear. We found that there were at least five different methods invented by the class of 30 students.

The students shared their methods with each other and found that they had all come to the same answer, although their methods were very different. Some students spent a great deal of time and many calculations while others took very little time. Those who found their answer very quickly typically used a 24-hour clock method while the others struggled with trying to work with a 12-hour clock. A very thoughtful discussion arose, as each student tried to defend his or her method to the others. Some had never thought of time in a 24-hour paradigm before and resisted the acceptance of the 24-hour model. The argument and discourse went on for some time, but finally the class came to a consensus about a method that was the most expedient and efficient. The beauty of the experience to the teacher and me was how the students' interest reached a level of discussion that left us almost completely out of the picture. They were thinking about each other's ideas and their own and comparing the efficacy of the methods used. In other words they were thinking about their thinking, comparing, making decisions, and deepening their understanding of the concept of *time* as it related to a problem they wanted to solve. I hasten to say that it works with adults too.

—Dick Konicek-Moran

As we all know, the ways in which schools sometimes operate make the time-consuming option described in the vignette above difficult to implement. Lisa Schneier sums the problem up succinctly in the following quotation:

> *The fact remains that schools are structured to bring students to fixed points of knowledge in a certain length of time. Teachers and students are accountable to elaborate structures of assessments that are wielding more and more power. These assessments carry with them assumptions about learning and knowledge that exert a constant narrowing force on the work of schools. Often the decision as it confronts teachers is whether to short-circuit substantive work that is happening in their classrooms in order to prepare students for these tests. How to balance these forces against the deeper knowledge that we want for students is a continuing question for me. (quoted in Duckworth 2001, p. 194)*

We have faith that since we are now poised on the cusp of a new era in teaching for conceptual understanding with the release and implementation of the *Next Generation Science Standards*, teachers can focus on fewer topics each year and teach for deeper understanding. With different means to assess student learning and the application of that learning, including continuous formative assessment, we can build a bridge from learner's initial theories about the way the natural world works and how science is practiced to where they need to be to understand scientific concepts and practices.

And now, in Chapter 2, we move to the history of science, to see how we may learn from the past so we can move forward in the present to prepare our students for a future that depends on a conceptual understanding of science and scientific practices.

## Questions for Personal Reflection or Group Discussion

1. Examine your own teaching practice. What percentage of an entire school year do you think you actually teach for conceptual understanding in science versus "covering the curriculum?" What initial change(s) could you make to shift that percentage more toward conceptual understanding?
2. The term *habits of practice* describes teaching practices that have become so routine that we don't bother to question them. Can you think of a habit of practice that interferes with teaching for conceptual understanding? What can you or others do to change that habit of practice?
3. Dick Konicek-Moran's NSTA Press series *Everyday Science Mysteries* and Page Keeley's *Uncovering Student Ideas* series are popular resources for uncovering what students (and teachers) really think related to scientific concepts. Think of a story or probe you may have used from one of their books that uncovered a lack of conceptual understanding. What surprised you about your students' (or

teachers') ideas? How did this chapter help you better understand why some students or teachers harbor ideas that are not consistent with scientific knowledge or ways of thinking?

4. Keep track of everyday or "sloppy use" of science terms for a designated time period as you find them in the media, in conversations with others, or even in your curriculum. Make a list and consider what could be done to change the way these terms are used in the public and school vernacular.
5. List some examples of concepts that you once may have thought you understood but later found you lacked clarity and depth of understanding.
6. Look at the list of crosscutting concepts on page 16. Review these concepts by reading pages 83–101 in *A Framework for K–12 Science Education* (NRC 2012) or online at *www.nap.edu/openbook.php?record_id=13165&page=83*. Identify examples of ways these concepts can be included in the curricular units you teach.
7. Change is more effective when learners experience it together, whether it is students learning a concept or teachers learning about teaching. How would you go about setting up a climate for intentional conceptual change within a community of learners at your school or organization?
8. React to Lisa Schneier's comments on page 22 regarding balancing time against deeper knowledge. How do you think the *Next Generation Science Standards* or your own set of state standards will fare against this issue of time for teaching versus depth of understanding?
9. Choose one "golden line" from this chapter (a sentence that really speaks to or resonates with you). Write this on a sentence strip and share it with others. Describe why you chose it.
10. What was the biggest "takeaway" from this chapter for you? What will you do or think about differently as a result?

## Extending Your Learning With NSTA Resources

1. Read and discuss this article, which shows how elementary children connect newly learned material to their existing knowledge: Kang, N., and C. Howren. 2004. Teaching for conceptual understanding. *Science and Children* 41 (9): 29–32.
2. Read and discuss this article, which explains how to create and use an interactive word wall: Jackson, J., and P. Narvaez. 2013. Interactive word walls. *Science and Children* 51 (1): 42–49.
3. Read and discuss this article, which describes how thought and language are intricately related: Varelas, M., C. Pappas, A. Barry, and A. O'Neill. 2001. Examining

language to capture scientific understandings: The case of the water cycle. *Science and Children* 38 (7): 26–29.

4. Read and discuss this article, which describes the crosscutting concepts: Duschl, R. 2012. The second dimension: Crosscutting concepts. *Science and Children* 49 (6): 10–14.
5. Read and discuss this article about use of the words *theory* and *hypothesis*: McLaughlin, J. 2006. A gentle reminder that a hypothesis is never proven correct, nor is a theory ever proven true. *Journal of College Science Teaching* 36 (1): 60–62.
6. Read and discuss this article about how word choice affects students' understanding of the nature of science: Schwartz, R. 2007. What's in a word? How word choice can develop (mis)conceptions about the nature of science. *Science Scope* 31 (2): 42–47.
7. Read and discuss this NSTA Press book about building data literacy: Bowen, M., and A. Bartley. 2013. *The basics of data literacy: Helping your students (and you!) make sense of data*. Arlington, VA: NSTA Press.
8. Read and discuss Chapter 3 "Foundational Knowledge and Conceptual Change" in Michaels, S., A. Shouse, and H. Schweingruber. 2008. *Ready, set, SCIENCE!* Washington, DC: National Academies Press.
9. The authors' NSTA Press series *Everyday Science Mysteries* (Konicek-Moran) and *Uncovering Student Ideas in Science* (Keeley) contain a wealth of information on children's alternative conceptions and strategies for eliciting children's ideas. Read and discuss sections from these books. You can learn more about these books and download sample chapters at the NSTA Science Store: *www.nsta.org/store*
10. Watch the NSTA archived *NGSS* webinar on developing and using models: *http://learningcenter.nsta.org/products/symposia_seminars/NGSS/webseminar6.aspx*
11. View videos of authors Dick Konicek-Moran and Page Keeley discussing the importance of understanding children's ideas: *www.nsta.org/publications/press/interviews.aspx*

# Chapter 2

## What Can We Learn About Conceptual Understanding by Examining the History of Science?

### History of Science in K–12 Science Education

Beginning in elementary school, children are fascinated by stories about scientists. Each year NSTA publishes a list of Outstanding Science Trade Books for Students K–12, which often includes books about great historical figures of science and their discoveries. For example, *I, Galileo,* by Bonnie Christensen, describes the many scientific contributions of Galileo and his controversial life during a time in history when pursuing scientific truth was at odds with the prevailing views of that time. The AAAS resources, *Benchmarks for Science Literacy* (2009), *Atlas of Science Literacy* volumes 1 (2001) and 2 (2007), and *Science for All Americans* (1988) all contain sections on the history of science for middle and high school grades, recognizing that certain episodes in the history of science can enhance the science curriculum and contribute to students' understanding of the scientific enterprise. Furthermore, *Science for All Americans* (AAAS 1988) gives two reasons why including some knowledge about the history of science is important, even for adults who do not go on to study science or be scientists: (1) generalizations about how the scientific enterprise works would be empty without concrete examples, and (2) some episodes in science are of great significance to our cultural heritage and understanding the development of thought in Western civilization. Even though the *Next Generation Science Standards* (NGSS Lead States 2013) do not include the history of science, this does not mean it is not important to embed historical perspectives into the curriculum (remember that standards are not a curriculum!). Learning about historical episodes and the scientists who advanced our understanding of the natural world over the past few centuries can certainly enhance and contribute to an understanding of the core disciplinary ideas of science while accentuating the nature of the human enterprise we call science.

### History of Science

There is a group of scholars who specialize in trying to understand the history, thinking, and philosophy of early scientists. These scholars read original research papers and try to understand how these people came to understand the world as they did. They also try to understand how and why ideas changed over time. Science is a way of solving mysteries about the natural world, and it has its share of dead ends as well as magnificent successes. Unfortunately, some school science textbooks often tell the story of science as though it were only the last chapter of a book.

### Aristotle

For centuries "natural history" followed the philosophy of Aristotle, a student of Plato and a teacher of Alexander the Great, who lived and worked in the third century BCE. His "scientific method" depended almost entirely on logic or dialectic. His methods were the standard of thinking through the medieval periods of history and it was not until the approach of the Renaissance and the work of Galileo and Descartes that new ideas built on the importance of experimentation became the new paradigm. Aristotle believed that the universe was composed of five elements: fire (heat), air (gas), earth (solids), water (liquids), and aether, or the heavenly substance that formed the universe of stars and heavenly bodies. He also believed that natural objects in the universe had their "natural places." Thus, rocks fell to earth and temperatures reached equilibrium because that was their natural place. Aristotle's views were held as the gold standard until the enlightenment of the 17th and 18th centuries in Europe and later in the American Colonies. However, there are scientists who still revere Aristotle, such as Armand Marie Leroi, author of *The Lagoon: How Aristotle Invented Science* (2014). Leroi believes that Aristotle's inquisitive nature made him the first scientist. Aristotle had not yet come up with the idea of experimental study, yet he had a certain curiosity about nature and did a great deal with taxonomy, the classification of living things. There are others who believe that Galileo was the first scientist because he insisted on experimental evidence to support his beliefs. Thus, we see another difference of opinion.

### Religion and Science

Historically, religions had a profound effect on people's beliefs, including beliefs about the natural world. People, most of whom were illiterate, were told what to believe and the Roman Catholic Bible was thought to be infallible and was taken literally. The population had little opportunity to read and make decisions for themselves. Universities were formed to train people to defend Church doctrines and combat what the Church considered heresy. It was into this society (the 15th to the 17th centuries) that early Western scientists including Galileo, Descartes, Copernicus, Kepler, and Brahe were born. It may be of interest to know that the Roman Catholic Church did not allow Copernicus's theory, written in 1510 AD, that considered the Sun as the center of the solar system, to be published until 1835. Of course the Muslims, Druids, Mayans, Chinese, Polynesians, and some lesser-known societies had developed astronomical theories and calendars in much earlier times. However, the methods of these societies were destroyed, lost, or discounted over time, and Western science predominated. As a modern-day example, it was only in the 1990s that Westerners were finally convinced that Polynesians possessed the scientific knowledge and skills to navigate the Pacific Ocean centuries ago to populate the Hawaiian Islands and reach *Rapa Nui* (Easter Island) centuries ago.

### *Modern Science*

Scientific activity is one of the main features of the contemporary world and, perhaps more than any other, distinguishes modern times from earlier centuries (AAAS 1988). The transition to modern-day science began with the scientific revolution, which took place in Europe during the late 16th and early 17th centuries. As a result, norms such as accurate recordkeeping, openness, and replication were established and scientists began to form communities that regulate and act as watchdogs over the members of that community. Science is a human endeavor and therefore it contains its list of scoundrels as well as its list of heroes (fortunately far more of the latter). As a discipline, it is founded on a set of principles and ethics, but that does not mean that everyone engaged in science follows those ethics all of the time. Sometimes, the pressure to get credit for being the first to publish a new idea or observation leads some scientists to withhold information or in rare cases, to even falsify their findings. There was a cold fusion claim in 1989 that supposedly showed atomic fusion and energy generated at room temperature but could never be duplicated. Scientists have been found to fit their data into their theories despite errors and discrepancies in their results (Brooks 2011). When such rare violations of the very nature of science are discovered, they are strongly condemned by the scientific community.

One particularly malicious fraud perpetrated on the scientific community was the Piltdown man. A prankster in England in 1912 buried fossil bones of an orangutan and a human, and provided scientists a reason for argument about evolution until 1953 when the prank was finally uncovered. However, these "science scoundrels" are the rare exception. The vast majority of scientists conduct themselves according to generally accepted ethical principles and rigorous adherence to standards for scientific work; for example, scientific claims must always pass the test of replication by other scientists. Thus we have many well-respected constructs from Isaac Newton, Charles Darwin, Marie Curie, Barbara McClintock, and many other "mothers and fathers of science" of the past as well as the present.

## Conceptual Change and the History of Science

Sometimes when we read historical accounts of science, we are surprised and even amused at how scientists could be so naive about certain concepts. Feeling superior to early scientists is a mistake because early conceptual ideas led the way to what we believe today. Without the minds and courage of early scientists' thinking, however wrong they may have been at the time, we would not have the theories that direct our present thinking. In the same way, early conceptual thinking by our students sometimes makes us prone to the "isn't that a cute idea" syndrome or the desire to "fix" students' thinking without recognizing the significance of their initial ideas.

We can learn a great deal about conceptual change from looking at the history of science. Philosopher Thomas Kuhn and Susan Carey, a cognitive psychologist from Harvard University—two important framers of today's understanding of science thinking—believe that there is a very close correlation between the changes in scientific thought over the centuries and those that we strive for in the classroom (Kuhn 1996; Carey 2009). Carey states in her book *The Origin of Concepts* that some changes in scientific theories have taken centuries to become accepted. So, it would not be hard to imagine that they may take years in the case of individual scientists or students engaged in restructuring personal knowledge.

Both Kuhn and Carey strongly emphasize the need for commensurability in conceptual change. *Commensurability* means that two theories can be gauged by the same common measure. However, in the history of science and conceptual change, *incommensurability* is more likely to be found, because people holding one conceptual scheme cannot find a common language by which to compare or contrast other schemes. In fact, Carey believes that incommensurability is *necessary* for conceptual change. By this, she means that real conceptual change occurs when two theories are so far apart that the people need to create a new and common set of structures to allow understanding and agreement:

> *[C]onceptual change is not the same as changing one's mind, acquiring new knowledge, or changing one's beliefs. Rather, it means creating new concepts not expressible in terms of previously available vocabulary. As I use the term, conceptual change requires incommensurability. ... Not all theory development involves conceptual change; often theories are merely enriched as new knowledge accumulates about the phenomena in the domain of the theory. Theory enrichment consists of the acquisition of new beliefs formulated over a constant conceptual repertoire. (Carey 2009, p. 364)*

Over the years, Kuhn softened his ideas on incommensurability somewhat (Kuhn 1996). A great deal depends on the conceptual ecology and the depth of emotional attachments of the communicants. In his later writings, Kuhn does not consider incommensurability to be as dramatic a barrier to theory change as previously stated. However, in some cases, the barriers of thought may be so far apart that it still may create a very difficult barrier to consensus.

### Joseph Black's Conceptual Revolution in Heat

In the mid-18th century, most scientists had not differentiated between *heat* and *temperature*, an important distinction. *Heat*, as the word was used back then, or what we would presently call *thermal energy*, is the internal energy of a substance produced by the motion of its atoms or molecules. The way we use the word *heat* today in a scientific sense refers to the transfer of thermal energy, such as through conduction, when two objects or systems are at different temperatures. *Temperature* is a measure of the average kinetic energy of the particles that make up matter. If you have equal volumes of hot water, one 30°C and one 40°C, and add them together, the total amount of thermal energy (commonly called *heat energy* in everyday language) is the sum of the thermal energy in both cups. However, to get the temperature, you would not add the two temperatures together. Energy from the warmer water (40°C) is transferred to the cooler water (30°C) until they reach an equal *temperature* of 35°C. In addition, *heat* is required to raise *temperature*. So, lumping *heat* and *temperature* together forms a sort of conceptual mess. But, early scientists did lump them together—*and* they also had the concept of *cold* as sort of an entity unto itself, which could produce changes in matter (i.e., water to ice). But there began to be problems with understanding certain phenomena that scientists were just beginning to be able to measure, creating an *incommensurability* between the current theory and the results that they were seeing in their experiments. Joseph Black, in his University of Glasgow lecture in 1872, demonstrated his theory about latent (energy needed to change a substance from one state to another, such as solid to liquid or liquid to gas) and specific heat (the amount of heat needed to raise the temperature of one gram of a substance one degree Celsius) that solved these discrepancies. He theorized that when one touches ice, heat *leaves* the hand; cold does not *enter* the hand. His theories of heat solved all sorts of problems. Carey states that after Black "in no theory … is cold conceptualized as a real entity, on par with heat" (2009, p. 374). In this particular case, the new concept was accepted in the community and eventually became the basis of the commonly accepted laws of thermodynamics, which also became vital to various engineering innovations, beginning with the steam engine. Scientists needed to talk with each other until they were speaking a language that all could understand, and *conceptual change* could occur. Today we still encounter conceptual difficulties with students when referring to heat and temperature that mirror early historical beliefs.

### *Science Textbooks and Current Science*

In another example of incommensurability, in the 1950s textbooks reported that the human cell contained 48 chromosomes (24 pairs). A later recount identified 46 as the correct number, and thus, it was changed in subsequent textbooks, and teachers changed their teaching. This was not conceptual change, just a correction of information. But this does point out that is important to realize that textbooks are often considered authorities, and in the best of circumstances (not that every school is able to enjoy the best of circumstances), they try to encapsulate the most current information ("Just the facts, ma'am!"). They do not tell the messy struggles involved in the obtaining of this information, and thus there is a danger that readers believe that science and science textbooks are infallible. Philosopher Alfred North Whitehead wrote that if a science forgets its heroes, it is lost (Kuhn 1996).

Science texts have not forgotten to mention science's heroes, but they do tend to leave out their errors and confusions. But in minimizing these, they tend to make students believe that making discoveries (changing ideas) is easy, and if it *is* so easy, make students wonder why it is so hard for them.

## Paradigms and Revolutions in Science

In his book *The Structure of Scientific Revolutions,* first published in 1962 and considered by many to be one of the most influential scientific books in the last 60 years, Thomas Kuhn introduced the idea of scientific *paradigms,* which we mentioned briefly in Chapter 1. A paradigm can be likened to a judicial decision in common law: "It is an object for further articulation and specification under new or more stringent conditions" (Kuhn 1996, p. 23). Examples of historic paradigms are Ptolemy's computations of planetary position, Lavoisier's application of the balance and introduction of oxygen, Maxwell's mathematization of the electromagnetic field, quantum mechanics, the theory of natural selection, plate tectonics, and of course, Newton's publication of the *Principia.* New paradigms, says Kuhn, provide new ways of looking at the puzzles in science that seem important but are unattainable under an old paradigm. He expands his explanation by stating, "Paradigms gain their status because they are more successful than their competitors in solving a few problems that the group of practitioners has come to recognize as acute. ..." (1996, p. 23). If the new paradigm increases the match between facts and predictions, and offers promise of future value to those scientists who follow that paradigm, more and more scientists may, over time, see its value and the nature of scientific research will change—just like the scientists who were able to finally understand the problems they were having with aspects of heat that Joseph Black explained with his theory.

*Normal science,* as defined by Kuhn, is the practice of scientists to attempt to "mop up" the questions that the new paradigm dictates or predicts. Scientists are not known for their willingness to adopt changes and give up their old ideas any more than students may be. Einstein was not happy with *certain* aspects of the quantum theory that indicated probabilistic results in the location of electron orbits. He is quoted as saying, "My God does not play with dice." (He could live with other parts of the theory, however, showing that people do not have to swallow a theory whole.) But quantum theory has allowed scientists to delve deeper into the atom and into energy puzzles that have led engineers to invent lasers, MRIs, diodes, and other practical devices for the use of humankind. New paradigms are often accompanied by new pieces of apparatus, and the building of improved telescopes, microscopes, and other equipment were necessary to test the new paradigms. Thus we have the building of the Large Hadron Collider in Switzerland to create and study the Higgs boson, a subatomic particle, which is vital to quantum theory.

In the early 18th century, compounds and mixtures and how they combined were not well understood. Two French chemists, Proust and Berthollet, debated this issue at length, but neither of their positions told exactly how two chemicals were able to combine to create a seemingly new substance (compounds) while others seemed to combine for a while, but could be returned to their old individual identities through physical means (mixtures). Chemists at the time were influenced by the Aristotelian and Newtonian principles of *affinity*: Things had their proper place and behaviors, and some things were drawn to others naturally. The chemists of the day suspected there was something they were missing but were not able to agree on what it was.

Enter John Dalton (1766–1844), a meteorologist, and therefore not held to the current chemists' paradigm. He was convinced that all substances were formed of small particles—atoms. He determined that certain atoms combined with other certain atoms in a manner proportional to their weight. He published his theory in the three-volume *A New System of Chemical Philosophy* (1808, 1810, 1827). This defined a *chemical reaction*. If two ingredients were not involved in a chemical reaction, then they remained intact as atoms and thus were a mixture. Finally, there was an answer to the problems the chemists were encountering and a useful differentiation into compounds and mixtures.

> *What chemists took from Dalton was not new experimental laws but a new way of practicing chemistry. … As a result, chemists came to live in a world where reactions behaved quite differently from the way they had before. It did take several generations for scientists using normal science to fit their experimental findings into this new paradigm. (Kuhn 1996, p. 134)*

## Author Vignette

I have in my possession a textbook written in 1845. It is written in a catechism style. There are questions—and answers to those questions—that were to be memorized and fed back to the teacher. (One question that amused me was, "What should one do in a lightning and thunderstorm?" Answer: "Draw one's bedstead to the middle of the room and commend oneself to the care of God.") In the section on chemistry, it asks why small bubbles tend to form around larger bubbles. The answer was given as "Because the larger bubble attracts them." Subsequent questions about attraction received the same answer. Only recently did I realize that this text was still operating under the Aristotelian principle of "affinity."

—Dick Konicek-Moran

## Similarity of Science Classrooms to Historic Communities

If classrooms are similar to historic communities, the arrival of a new way of looking at phenomena might very well need a new theory that offers better and believable results rather than merely showing students that an old theory no longer works. Kuhn also gives us a clue about how we might enter into conversations with our students when trying to initiate conceptual change:

> *[W]hat the participants in a communication breakdown can do is recognize each other as members of different language communities and then become translators. ... Each may, that is, try to discover what the other would see and say when presented with a stimulus to which his own verbal response would be different. If they can sufficiently refrain from explaining anomalous behavior as the consequence of mere error or madness, they may in time become very good predictors of each other's behavior. Each will have learned to translate the other's theory and its consequences into his own language and simultaneously to describe in his language the world to which that theory applies. (1996, p. 302)*

But we must be careful here in defining what it means to become a "translator." Kuhn does not mean that we may gain understanding by merely substituting one word for another—the words in our languages may not have the same meaning. Imagine substituting "Bruce Wayne" for "Batman" in a story. It would alter the meaning while maintaining the reference. So by translation, we mean that the "core" of the theory must be translated into the thought processes of each of the communicants. Sometimes a common language can be discovered; other times not. Perhaps, as Carey suggests, we should substitute the word "interpret" for the word "translate" (2009, p. 369).

## Case Study on an Examination of Heat in the Classroom

Let's look at a typical situation in a middle school classroom. Fouad, the teacher, suspects that many of his students do not distinguish between the concepts of *heat*, *thermal energy*, and *temperature*. Furthermore, they come to middle school using the word *heat* in the way it is commonly used in our everyday language. The term *thermal energy* is an unfamiliar term to most children prior to learning about it in school. The word *heat* in science actually refers to energy in transit. When we refer to the internal energy of a system, we are actually referring to *thermal energy*, commonly called *heat energy* in familiar language. One of the changes made to the *Benchmarks for Science Literacy* is when they transitioned from using the term *heat energy* to *thermal energy* in the language of the middle school learning goals (AAAS 2009). Currently the *NGSS* introduce the term *thermal energy* at the middle school level to describe the internal energy of a substance produced by the motion of its atoms or molecules. It is the sum of the energy associated with the kinetic and potential energy of each particle in an object, material, or substance. However, many teachers tend to use the familiar term *heat energy* until students are ready to distinguish between heat as energy in transit and thermal energy as internal energy. Regardless of the terminology used, and how vexing the terminology can be when used in so many different ways, the conceptual understanding is what is important.

Fouad may have given the class a formative assessment probe such as "Mixing Water," (Figure 2.1) from *Uncovering Student Ideas in Science, Volume 2* (Keeley, Eberle, and Tugel 2007) or had them use the stories, "Cooling Off" or "Party Meltdown" from *Yet More Everyday Science Mysteries*, as discussion starters (Konicek-Moran 2011). Both "Mixing Water" and "Cooling Off" ask students to predict the resulting temperature after mixing equal amounts of water at different temperatures. For example, if the water in one of the glasses was 50°C and an equal amount of water in another glass was 10°C, what would the temperature of the water in the resulting combination be?

**Figure 2.1. "Mixing Water" Probe**

## Mixing Water

Melinda filled two glasses of equal size half-full with water. The water in one glass was 50 degrees Celsius. The water in the other glass was 10 degrees Celsius. She poured one glass into the other, stirred the liquid, and measured the temperature of the full glass of water.

What do you think the temperature of the full glass of water will be after the water is mixed? Circle your prediction.

**A** 20 degrees Celsius
**B** 30 degrees Celsius
**C** 40 degrees Celsius
**D** 50 degrees Celsius
**E** 60 degrees Celsius

Explain your thinking. Describe the "rule" or reasoning you used for your answer.

*Source:* Keeley, Eberle, and Tugel 2007.

Should it become evident that students are adding the temperatures of the water in each glass, Fouad can assume that students are looking at temperature and heat from an additive conceptual view and may think that heat and temperature are the same. As mentioned previously, the total amount of *thermal energy* in the combined cup is the sum of the *thermal energy* in each of the two cups. But the *temperature*, the measure of the *average* kinetic energy, of the mixed glasses would be approximately 30°C (in actuality, it may be slightly less because a small amount of heat may be lost to the surrounding environment). When the cooler water and the warmer water are mixed together, a transfer of energy occurs between particles when they come in contact with each other. Heat moves from the molecules in the warmer water to the molecules in the cooler water until they have the same average energy (temperature). Energy is transferred until a thermal *equilibrium* is reached.

An understanding of the particulate nature of matter as well as the kinetic molecular theory of matter is necessary to understand thermodynamics thoroughly. However, it is possible to distinguish between *heat energy* (thermal energy) and *temperature* at a qualitative level through a simple demonstration. Fouad displays a beaker of water that has a very high temperature—near boiling, verified by using a thermometer. He asks the students if they would be willing to immerse their hand in the beaker. The answer is almost always an emphatic, "NO!" He then takes a small drop of the water and dramatically lets it fall on his hand with no obvious discomfort. Fouad then asks the students to discuss what they have seen and to try to explain their observations. He tries to lead the discussion toward the differences in amounts of water and the thermal energy they contained, even though both the droplet of water and the water in the beaker were the same temperature. As Fouad listens to the students' comments, he will hear their theories about heat or *heat energy* (using this term as a stepping stone to introducing the term *thermal energy*) and *temperature*. This is a type of formative assessment embedded in the demonstration. Most middle school students will realize that the amount of thermal energy contained in the water in the beaker was far more than that in the droplet and that temperature, even though it was the same, is different than the amount of thermal energy contained in each water sample. Thus, as Joseph Black concluded back in the 1800s, *temperature* and *heat* (which was the term they used for thermal energy back then) are two entirely different concepts.

The following anecdote from the history of science parallels the thinking many modern-day students have about *heat* (thermal energy) and *temperature*. In the 17th century, even before Joseph Black, there was a group of scientists in Florence, Italy, some of whom were students of Galileo. They were commissioned by the local authorities to experiment in the area of heat. They called themselves the *Experimenters* or the *Accademia del Cimento* and saw as their goal the breaking away from Aristotelian logic as a way of practicing science. (Remember, it was only after Galileo's famous set of trials with the inclined plane and inertia that Western science turned to "proving" truth with experiments and not just philosophical logic.) They dabbled in all sorts of heat-related experiments that were driven by what Wiser and Carey (1983) called the "source recipient" theory: Something hot or cold (say, a fire, or an ice cube) was the *source,* and any nearby mass was the *recipient* of the heat or cold. In other words, ice released cold into water and cooled it, and hot objects warmed nearby objects. It is ironic that although they were attempting to dispel the ideas of Aristotle, their theory required a belief in Aristotle's natural state. In essence, cold or hot objects "tended" to surrender their heat or cold to surrounding matter (Wiser and Carey 1983). They invented a thermoscope to measure differences in what they called the "degree of heat or cold." The device included a vessel of alcohol and a tube into which the alcohol expanded or contracted.

The experimenters did not reach any notion of thermal equilibrium (the idea that heat energy will flow to a cooler area) because of their conceptual limitations. One particular experiment stands out to show how a group or an individual can hold two contradictory ideas at the same time, just as our students may. The Florentine experimenters developed a chemical reaction in a metal box that produced heat. They noticed this box would not melt paraffin. Then, they put a block of metal on a fire until it was hot enough to melt paraffin. They deduced that the metal block had more degrees of heat than the box holding the chemical reaction. They then put both the box and the block in ice and found that the chemical reaction box melted more ice than the block during the same amount of time. They then deduced that the box with the chemical reaction had more degrees of "heat" than the block. How? They noticed the contradiction but were at a loss to explain the different kinds of "heat" without separating the concepts of *heat* and *temperature*. (The melting point of paraffin is a function of temperature but the melting of ice depends on the amount of heat provided.) It is interesting to note that the source-recipient theory works very well with the exception of certain phenomena. Thus, agreeing to a separation of the concepts of *heat* and *temperature* and admitting the existence of molecules in motion as a source of thermal energy and heat transfer would mean giving up the existence of *cold*. Marianne Wiser showed in her study that adolescents clung to theories that were very close to those held by the Florentine experimenters (Wiser 1988).

## Correlations Between Historic and Classroom Persistence of Theories

There are direct correlations between humans holding on to their idiosyncratic concepts and that of scientists doing the same in the past. Charles Darwin's affection for his theory of *pangenesis* was so strong that he refused to look at any other alternative, despite evidence to the contrary. In the 19th century, one of the biggest puzzles was that of explaining heredity. Many breeders had tried and succeeded in creating hybrids and had no idea how characteristics were passed from parent to offspring. Darwin, publishing in 1868 *The Variation of Animals and Plants Under Domestication*, was practically forced to come up with an explanation of inheritance. He did so in his theory of *pangenesis*. This theory suggested that traits were passed on to offspring through *gemmules* that were shed from the body cells of the mother and father and accumulated in the reproductive organs prior to fertilization. Despite criticisms from his loyal colleagues Hooker and Huxley, he defended his theory as the best there was to offer at that time. It seems doubtful that even if Darwin had access to Mendel's data on the ratios of inheritance and his findings about dominant and recessive traits (Mendel's paper on peas lay fallow in some dusty repository at the time), Darwin would have been swayed.

Darwin's cousin Francis Galton attempted to confirm *pangenesis* by introducing blood into certain rabbits and hoping for variations in offspring. His hypothesis stated that the *gemmules* for heredity would be carried in the blood. Darwin even participated by giving advice. Then when Galton disputed the theory because of his results, Darwin defensively criticized the study, even though he had participated! So we can see that even scientists who were paradigm breakers themselves are sometimes unwilling to give up their favorite theories unless forced.

Mario Livio, in his book *Brilliant Blunders*, makes a strong case for the importance of accidents in the pursuit of knowledge in science. It is through these mistakes that new knowledge is created. Some blunders of our scientific geniuses led to even greater discoveries. However, he stresses that scientific theories have no absolute or permanent value:

> *As experimental and observational methods and tools improve, theories can be refuted, or they may metamorphose into new forms that incorporate some of the earlier ideas. ... Darwin's theory for the evolution of life by means of natural selection was only strengthened through the application of modern genetics. Newton's theory of gravity continues to live as a limiting case within the framework of general relativity. (Livio 2013, p. 269)*

As we look at the history of science and at the words of science historians, should we not heed their findings and encourage our students to take a risk and feel free to make mistakes as they seek to understand our universe? The history of science is replete with risks taken by men and women who followed their beliefs and pitted them against the natural world and their fellow colleagues. Many of our students are unwilling to be "wrong" when they are engaged in science. We need them to open their minds and to communicate with others about their understandings. It is only then that they can be free to listen, consider, and revise their thinking toward a greater consensus, as the scientific communities have done over the centuries.

Two people have done an amazing job of helping us to understand the nature of science. We are speaking of the late Carl Sagan and his protégé Neil deGrasse Tyson, each of whom is responsible for developing a program series shown on TV and available on DVD, called *Cosmos*. Sagan's programs were produced in the 1980s while Tyson's is current. These programs not only explain the supposedly deep secrets of astrophysics in an understandable and uncomplicated manner, but the messages about scientists in history makes these historical characters live as real people who overcame great difficulty to bring scientific knowledge where it is today. We highly recommend watching it yourself and using it in your classroom as well. You can find more information online about these men and the *Cosmos* program.

To sum up, imagine a teacher asking a class of middle school students, "When I put ice in my glass of warm lemonade, is cold moving from the ice to the lemonade or is heat moving from the lemonade to the ice?" The discussion that would arise would be most interesting and informative. Questions like this can lead to learning if we accept that often the ideas of our students were once the same ideas held by scientists in centuries past.

## Questions for Personal Reflection or Group Discussion

1. Think of a historical episode in science that you learned about when you were in school or that you read about. How did that historical episode contribute to your understanding of science?

2. How do you relate changes in scientific knowledge throughout history to modern-day changes in belief systems?
3. How do you think case studies from the history of science might be used in the science classroom?
4. Share some examples of incommensurability you have experienced, either as a student or a teacher.
5. How does your textbook (or other instructional materials) portray historical episodes in science?
6. What does the word *paradigm* mean to you? What are some other examples of paradigm shifts that were made in science throughout history?
7. Search for the meaning of the word *Einstellung* and discuss its meaning with your colleagues. How does this word relate to scientific thinking prior to modern-day science? How does it relate to your teaching and your students' learning of science?
8. Why do you think John Dalton's being a meteorologist allowed him to "think outside the box" when it came to a chemistry problem?
9. Compare historical scientific communities with classroom scientific communities.
10. Why do you think Darwin might not have been swayed in his theory of *pangenesis* by knowledge of Mendel's work in genetics?
11. Choose one "golden line" from this chapter (a sentence that really speaks to or resonates with you). Write this on a sentence strip and share it with others. Explain why you chose it.
12. What was the biggest "takeaway" from this chapter for you? What will you do or think about differently after reading this chapter?

## Extending Your Learning With NSTA Resources

1. Check out the NSTA annual list of Outstanding Science Trade Books for Students K–12 at *www.nsta.org/publications/ostb*. Look for books that feature historical accounts of science and scientists.
2. *Science for All Americans*, Chapter 10, provides an eloquent and detailed account of what science-literate adults should know about great historical episodes in science. You can read it here: *http://www.project2061.org/publications/sfaa/online/chap10.htm*
3. Chapter 10 of the *Benchmarks for Science Literacy* includes a section on historical episodes in science. The essays that accompany the benchmarks provide insight into what effective curriculum and instruction entail while integrating historical

perspectives into the content students are learning. You can read it here: *www.project2061.org/publications/bsl/online/index.php?chapter=10*

4. The AAAS/Project 2061 Strand Maps on Historical Perspectives in Science can be accessed through the National Science Digital Library at *http://strandmaps.nsdl.org/?chapter=SMS-CHP-1094*. There are nine maps on this site: The Copernican Revolution, Classical Mechanics, The Chemical Revolution, Relativity, Moving the Continents, Splitting the Atom, Explaining Evolution, Discovering Germs, and The Industrial Revolution.

5. Read the teacher background notes for the probe "Mixing Water" (Keeley, Eberle, and Tugel 2007). Should you want more information on this thermodynamics topic, we recommend you read the teacher notes in the *Everyday Science Mysteries* stories "Cooling Off" or "Party Meltdown" (Konicek-Moran 2011, 2013b).

6. If you are still puzzled by *heat, thermal energy*, and *temperature*, try out the "Science Objects" *Energy, Thermal Energy, Heat*, and *Temperature* at *http://learningcenter.nsta.org/search.aspx?action=quicksearch&text=science%20object*

7. Check out Joy Hakim's "The Story of Science" books for historical accounts of science that include Aristotle, Newton, and Einstein. You can find information about these books at *www.nsta.org/store/search.aspx?action=quicksearch&text=Joy%20Hakim*

8. Visit the NSTA website for the archives of NSTA journals *Science and Children, Science Scope*, and *The Science Teacher*. In the search box, enter "History of Science" and you will find several informative articles.

   - *Science and Children*: *www.nsta.org/elementaryschool*
   - *Science Scope*: *www.nsta.org/middleschool*
   - *The Science Teacher*: *www.nsta.org/highschool*

# Chapter 3

## What Is the Nature of Science, and What Does It Mean for Conceptual Understanding?

The National Science Teachers Association (NSTA) has a position statement on its website on the nature of science. The first line in the preamble to the position statement reads, "All those involved with science teaching and learning should have a common, accurate view of the nature of science" (NSTA Board of Directors 2000). When the public reviews and the focus group reviews of the early drafts of the *Next Generation Science Standards* (*NGSS*) were examined, science educators expressed concern that the nature of science was not included in the *NGSS*. As a result of this feedback, changes were made to the *NGSS* that highlighted nature of science concepts throughout the document. In this chapter, we will examine the *nature* of science and the connection to conceptually understanding science.

Why is the nature of science important to science educators? One of the most telling reasons comes from our goals to develop a public understanding of science. In a society where citizens must make informed decisions about problems that confront them, understanding both scientific knowledge and how that knowledge was obtained is important. Scientists bring information, insights, and analytical skills to bear on matters of public concern (AAAS 1988). Often they can help the public and decision makers to understand the likely causes of natural and technological events and to estimate the possible effects of proposed policies or actions. There have been instances when political, economic, or societal interests have disregarded or modified the scientific information released to the public, but the public was also privy to other information upon which they could cast their judgment. To be an informed citizen in today's society, it is imperative that the nature of science be part of every student's K–12 science education. This recognition of the importance of developing understanding of the nature of science led to its inclusion in the *NGSS* as a set of eight understandings that are incorporated into the *NGSS* practices and crosscutting concepts. Additionally, many state standards include learning goals related to the nature of science.

In *Young People's Images in Science*, a book based on years of research on children's notion of the nature of science, the authors agree that there are three aspects of science that are necessary for the public understanding of science: (1) an understanding of some aspects of science content, (2) an understanding about the scientific approach to inquiry, and (3) an understanding of science as a social enterprise (Driver et al. 1996, p. 13). Similarly, *Science for All Americans* (AAAS 1988) and *Benchmarks for Science Literacy* (AAAS 2009) point out three areas of the nature of science important for developing science literacy: (1) how scientists view their work (The Scientific World View); (2) how scientists go about their work (Scientific Inquiry); and (3) individual, social, and institutional activities of science (The Scientific Enterprise).

Chapter 1 of this book commented on the lack of understanding of terms used in science, such as *theory*, *hypothesis*, and *experiment*. In addition to using correct terminology it is also important that students be aware of the epistemology involved in science. Science is a social enterprise, and learning about its structure and its practitioners is a necessary part of science education.

## What Is the Nature of Science?

Science can and has been defined in many different ways, but the common attributes remain much the same:

- Science is more than a collection of facts.
- Science knowledge is made up of current understandings of our natural systems and is subject to change.
- The body of knowledge is constantly being revised, extended, and refined.

Some researchers and science historians believe that there are at least three different models to describe the way science works: Logic and Reasoning, Theory Change, and Participation in Science Societies (NRC 2007).

### *Logic and Reasoning*

This model suggests that science works on a basis of logic and argumentation about evidence. The followers of Bärbel Inhelder and Jean Piaget espouse this view. It promotes an image of scientist-as-reasoner and problem solver. This means the scientist solves problems through logical strategies that can be related to all subject areas in science. In other words, if we teach students logical thinking and problem solving, they can go on to be scientists in any scientific domain. Once there, they just have to become acquainted with the facts and evidence in that domain (e.g., zoology, chemistry, or archeology) in order to become good scientists. Researchers who champion this view of science are Columbia's Teachers College educational psychologist Deanna Kuhn (1995), New Zealand educational psychologist Ted Ruffman and colleagues (1993), and University of California, Davis, psychologist, Zhe Chen along with developmental psychologist at Carnegie Mellon, David Klahr (1999).

### *Theory Change*

The second model can best be shown by the work of science philosopher Thomas Kuhn and Harvard's cognitive psychologist Susan Carey, both mentioned in Chapter 2. This model supposes that it is historical events and the philosophical constructs of society both within and without specific scientific domains that lead science. Proponents of this model would see science developing as a result of conceptual or theory change. There is a paradigm shift that every so often promotes a revolution in the way scientists practice their work and see the world. Examples would include the Galilean, Newtonian, and Einsteinian shifts (we have discussed each of these paradigm shifts in greater depth in other parts of this book). They also emphasize the importance of incommensurability among scientists that often makes the shifts happen slowly; and they believe this has relevance to the classroom, where both incommensurability and conceptual shift are necessary components for conceptual change.

### *Participation in Science Societies*

Many anthropologists, ethnologists, social psychologists, and cognitive and developmental psychologists view science as a collection of practices that include networks of participants and institutions. This model of science emphasizes the importance of society in the growth of scientific knowledge. The model goes beyond evidence-based reasoning but does not negate it. This means that scientists must not only have skills in logical reasoning and awareness of the facts and evidence of the scientific domain (e.g., genetics or physics) but also communication skills in order for them to negotiate their ways through the norms of the scientific society (NRC 2007).

## Substituting Science Practices for the Scientific Method

A practicing scientist has said that the practice of science is like searching for a black cat in a dark room when there may not even be a cat in the room (Firestein 2012). Science is not a step-by-step recipe for discovery; nor is it a methodical, systematic way of investigating the natural world that is rigidly followed in all areas of science. Rather, science is a human endeavor that sometimes even involves bumbling about in that "dark room" of unknowing. This belies what we have been taught about science over the ages and span of K–12 education. We've been brought up to believe in a formulaic "scientific method" developed by Sir Francis Bacon and Rene Descartes in the 17th century, which includes stating a question, formulating a hypothesis, carrying out an experiment, analyzing data, then drawing a conclusion that is then communicated to the community. Although there is a safe uniformity to this method, it is not what happens consistently in the scientific community, although it is often used as a template for reporting scientific efforts.

Scientists often use the term *scientific method*, but they understand its meaning in the context of their scientific discipline. "There simply is no fixed set of steps that all scientists always follow, no one path that leads them unerringly to scientific knowledge" (AAAS

1988, p. 4). To students, however, *scientific method* implies that there is a definite sequence of steps that all scientists follow. Merely changing our language to refer to *a* scientific method, rather than *the* scientific method changes the conceptual meaning. Figure 3.1 is an example of a formative assessment probe that reveals students' common interpretations of how scientists conduct investigations.

**Figure 3.1. "Doing Science" Probe**

**Doing Science**

Four students were having a discussion about how scientists do their work. This is what they said:

Antoine: "I think scientists just try out different things until something works."

Tamara: "I think there is a definite set of steps all scientists follow called the scientific method."

Marcos: "I think scientists use different methods depending on their question."

Avery: "I think scientists use different methods but they all involve doing experiments."

Which student do you most agree with? ____________________

Explain why you agree with that student and include why you disagree with the other students.

*Source:* Keeley, Eberle, and Dorsey 2008.

## Author Vignette

I developed the "Doing Science" formative assessment probe to reveal the ideas students have about how scientists do their work. In the K–12 professional development I provide, I often use this probe to elicit teachers' commonly held ideas about the "scientific method." In a recent workshop for 85 K–12 teachers, I used the anonymous elicitation technique, Sticky Bars (Keeley 2008), to display the data on the teachers' responses to the "Doing Science" probe. Each teacher recorded an answer choice on a sticky note. The sticky notes were then collected and a bar graph of overlapping sticky notes was created from the teachers' responses.

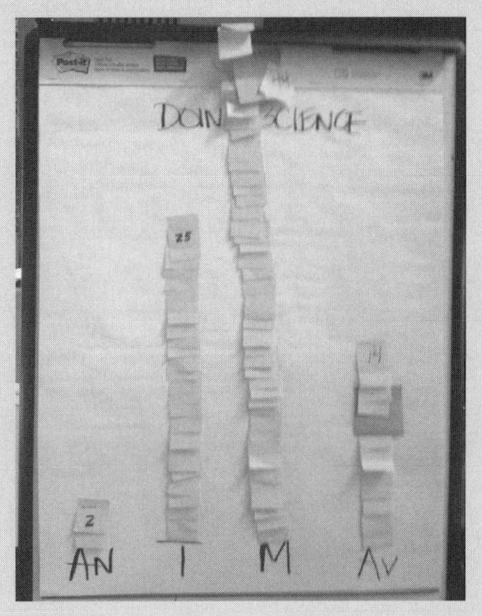


Two teachers (2%) selected Antoine: "I think scientists just try out different things until something works." Twenty-five teachers (29%) selected Tamara: "I think there is a definite set of steps all scientists follow called the scientific method." Forty-four teachers (52%) selected Marcos: "I think scientists use different methods, depending on their question." Finally, 14 teachers (17%) selected Avery: "I think scientists use different methods but they all involve doing experiments." As we discussed the responses, it was clear to the teachers that all their years as K–12 students of science, coupled with the materials they use to teach K–12 science, gave them a very limited view of the way science is conducted. Teachers who realized this after they chose Tamara said they knew that science is conducted in a variety of ways, including experiments, modeling, long- and short-term observations, field studies, collections, and so on. However, they explained that the "scientific method" was instilled in them from their own education, textbooks, posters on their walls, science fairs, and their state standards. Even though they knew science is conducted in a variety of ways, they still fell back on their common misconception. They acknowledged that it was no wonder that their students saw the scientific method as the one and only way to do science. The teachers acknowledged that they needed to be explicit in helping students understand science is conducted in a variety of ways, depending on the question and the branch of science. They also expressed concern that science fairs and the format for writing up lab experiments may be contributing to the common notion of a fixed set of steps all scientists follow. We concluded that it might be better to refer to *a* scientific method, rather than *the* scientific method, explaining to students that science is usually methodical but can take many different approaches.

—Page Keeley

Both *A Framework for K–12 Science Education* (NRC 2012) and the *Next Generation Science Standards* (NGSS Lead States 2013) avoid the use of *scientific method* and instead focus on using *scientific practices* in the language:

> *Seeing science as a set of practices shows that theory development, reasoning, and testing are components of a larger ensemble of activities that includes networks of participants and institutions, specialized ways of talking and writing, the development of models to represent systems or phenomena, the making of predictive inferences, construction of appropriate instrumentation, and testing of hypotheses by experiment or observation. (NRC 2012, p. 43)*

## Driven by "Ignorance"

Columbia University biologist Stuart Firestein, in his book *Ignorance: How It Drives Science*, shows how science moves forward by searching for ignorance: the things we don't know. *Ignorance* often has a pejorative connotation, but it is our admitting what we do not know and then searching for the answers that makes science go. (Charles Darwin warned us that we had best perceive our ignorance clearly!) What are really important in science are questions. A scientist decides which part of the dark room of ignorance he or she wants to explore and then asks good questions. One may find out that there is no "black cat" there and have to start over in another part of the room or in another room entirely, even after years of research. Kepler spent six years trying to find why there was a tiny error in the predictions of the Martian orbit only to ultimately discover that the planets in our solar system revolved around the Sun in elliptical orbits and not circular ones. This frustrated many astronomers at that time because ellipses were not seen as perfect as circles and perfection was expected in God's world. As a result, Kepler's findings were not accepted for a long time (Firestein 2012). But, without Kepler's work it is possible that Newton might not have accomplished what he did; Newton stated that if he had seen further than others it was because he had stood on the shoulders of giants. Newton was not a humble man, yet even he knew that prior ideas were important and lead to new and innovative theories.

The history of science contains many stories of scientists who have spent decades on a question only to find that they had reached a dead end. There have also been scientists who skip from question to question getting important results and scientists who have worked for decades on just one question, producing new evidence and, of course, many new questions.

What this means for teachers is that it is important to applaud uncertainty in the classroom. We should acknowledge there are many things we don't know and celebrate this (perhaps by making a bulletin board of "Things We Want to Explore"), and create an atmosphere of discovery in our classrooms. We also should find ways to praise *persistence* in pursuing answers to questions.

## Creating the Prepared Mind

Many bacteriologists saw the same phenomenon that Alexander Fleming saw in 1928, where a mold named *Penicillium notatum* delivered a death blow to staphylococcus bacteria on an agar plate. Fleming actually viewed this as an interesting puzzle to be probed, whereas others before him merely cursed their bad luck of having an experiment ruined and threw the whole thing out. Seeing puzzles in our results often result in discoveries beyond our wildest hopes. Enrico Fermi, the noted Nobel physicist, is said to have often told his students that, "an experiment that successfully proves a hypothesis is a measurement; one that doesn't is a discovery. A discovery, an uncovering—of new ignorance" (Firestein 2012, p. 57).

Fleming's discovery of penicillin is often attributed to a term, *serendipity*, coined in 1754, meaning "finding something of value that you were not looking for." Many items that we find useful today were discovered through serendipity, including Corn Flakes, vulcanized rubber, inkjet printers, Silly Putty, the Slinky, and chocolate chip cookies. Radioactivity, x-rays, and infrared radiation were all found by researchers who were looking for something else.

Perhaps Fleming's mindset had a great deal more to do with the discovery of the value of penicillin than mere serendipity. "Chance favors the prepared mind," said Louis Pasteur, and perhaps Fleming's mind was more prepared to see the value of the phenomenon than others. It's important that we think about how we help students, and or for that matter, teachers, to develop a "prepared mind."

## Language and Social Constructs

Although this may sound counterintuitive, science is not found in the phenomena being studied. Rather, it is found in the *constructs* or *ideas* about those phenomena that a community of scientists has socially negotiated in order to interpret and understand them (Millar, Driver, Leach, and Scott 1993). For example, the study of zoology is not merely about animals, but rather it is the (1) beliefs and understanding encompassed in (2) a specific language written about animals that zoologists have developed since the origination of this particular scientific domain. These beliefs and understandings, and even the language used, influences new beliefs and possible study directions. This does not mean that there is no empirical (or evidence-based) foundation for this knowledge. There are scientists such as Jane Goodall who spent 45 years in the wilderness studying chimpanzees and collecting data about the behavior of these mammals. Goodall spent months in Tanzania just waiting for the chimps to accept her presence before she could begin to observe and collect data. She was endowed with this kind of patience from childhood when, at the age of four, she spent hours in a henhouse watching to witness how chickens laid eggs while her parents reported her to the police as missing (Goodall 1985). Goodall realized that what she was observing challenged our conventional knowledge of the behavior of chimpanzees. Where researchers prescribed to the notion of "primitive" chimpanzees as vegetarians living simply, Goodall found highly intelligent, emotional creatures with distinct personalities living in complex social groups, using tools to solve problems, and even eating meat. Goodall's work shattered two long-standing myths: the idea that only humans could make and use tools and the belief that chimps were passive vegetarians. But Goodall had to be aware of the conventions and language used by her fellow researchers in order to be able to communicate her (then) revolutionary ideas to them.

## Author Vignette

I had the good fortune to meet Jane Goodall in the 1980s in Dar es Salaam, and found her to be a most sweet and humble person. After we attended her presentation, she asked a small group of us to look over her original notes to find relevant data for her new book that targeted how offspring of the original generation assumed their roles in parenting.

As I perused her diaries, I was amazed at the detail of her observations and remained amazed later when I saw how every observation she made was transferred into a coded language she was using decades later to answer questions about childhood-parenthood rituals of her original subjects. Today she continues her work for the creation of protected spaces for wild animals. She has written five major books based on her research and is considered to be one of the most revered and respected scientists in her field.

—Dick Konicek-Moran

As Michael Brooks, a British physicist and science writer, states: "The onus (to prove one's findings) is on the innovator. … Scientists can't be reeds blown this way and that by every new fad. Science is a battleground. It is written into the constitution of science that the road to Stockholm will be lined with jeering colleagues" (2011, p. 95). Goodall was an experienced scientist and kept her diaries for future use. One difference between an experienced scientist and a novice is he or she realizes that data from one investigation may hold both questions and answers for investigations yet to be attempted. Today, this can be mirrored in students' science notebooks, where they keep a record of all data and observations collected, which can be useful in later investigations.

## The Importance of Questions

It has almost become a cliché to say that when one tries to answer a question, 10 more will probably pop up. It is often those 10 questions that change the outlook of science and create a new paradigm. Nobel physicist Isadore Issac Rabi claimed that he owed his mother his success because she didn't ask what he did in school that day. Instead, when he came home from school she asked him, "Izzy, did you ask a good question today?" We need to help our students to ask good questions, ones that are personally theirs and lead to their sustained interest in science and persistence in finding answers and solving puzzles.

Where do these "good questions" come from? Why, from the students themselves, if they are encouraged to ask them. We should be amazed by and envious of our younger students' ability to view the world with wonder and awe. We used to experience this wonder and awe, but as adults, we often find it more difficult to use our imaginations and thus make contact with our own imaginative responses as well as those of our students (Vygotsky 1962). If we can regain this, we might be able to overcome our inability to sometimes think like a child and connect with children in a meaningful way on their level.

Karen Gallas, in her book *Talking Their Way Into Science* (1995), writes, "Each child has boundless questions that, once loosed, proliferate almost uncontrollably. Further, when the questions are valued and honored in a child's classroom community the child's desire to know becomes translated into an identification with the subject he or she questions" (p. 98). Karen has been teaching first and second grade for 40 years, with a brief stint as a faculty member at the University of Maine. Every day, she allows her children to engage in a "science talk" session, where she has taught them to enter into discourse about their own questions with civility and purpose. Her curriculum develops from the things she learns from observing these sessions—in which she is a participant and filter, not an authority figure.

## Author Vignette

I used a particular activity about the nature of doing science in my classes for over 30 years. It's called, "What's in the Bag?"

Approximately 20 brown paper bags were numbered 1–5 (four of each number) and into them were placed objects that I thought were familiar to the students. One set of each numbered bag contained different objects, but not necessarily the same objects (i.e., the #1 bags didn't all contain the same objects). All bags were stapled shut.

My students tried to discover what was in the bags, using any technique they wished *except* opening the bag. A colleague was recruited to put objects in the bags so that I did not know what was in the bags either. It not only made teaching more fun for me but also added to the idea of the mystery of scientific endeavors that the exercise was trying to get students to understand because even I, as the "authority figure," wasn't totally in control of the process.

After each group constructed their opinions of what was in the bags, I asked them if they had made any assumptions before beginning the activity. They most often answered that they assumed that all of the bags numbered alike would have the same objects in them. (Thus, some teams did not look at all of the bags of each number.) They also assumed that there would be nothing alive, dangerous, or completely esoteric in the bags. They were correct about the last three but not about the numbering assumption. *Lesson 1:* It is difficult to enter into an inquiry without making some assumptions, which may influence your methods or your observations.

When they had listed the items they thought were in the bag, we also listed all of the processes they had gone through to arrive at their conclusions: using their senses (observing), measuring, comparing to objects known to them, hypothesizing ("I think it is this because …"), experimenting, and drawing inferences. *Lesson 2:* We use many of the usual scientific techniques even to answer a simple question like this.

Now came the moment of truth—or so they thought. "Let's open the bags and see if we are right," they would say. When I collected the intact bags, there was almost always a shout of outrage. Some were so involved in the activity that they became emotional. I would ask them why they thought I wanted the bags to stay sealed, and eventually some students would answer that real scientists could not "open the bag" to verify the answers that they had been trying to solve. They had to go to the next step with the data and evidence they had, even if they weren't perfectly sure of everything. *Lesson 3:* Science and scientists live in uncertainty.

—Dick Konicek-Moran

## Science Teaching and the Nature of Science

We believe that science teaching might be in the midst of a paradigm change right now. Current research has shown us new ideas about how and when students learn best. Social media is rampant and information on just about anything you want to know is at your fingertips. The *NGSS* are combining content (disciplinary core ideas) with practices and crosscutting concepts rather than addressing them as separate standards; and science equipment, instructional materials, and new media are making it possible to do things with students we would have never imagined a decade ago.

Prior to the 1980s, we looked at a classroom full of children as a sea of minds to be filled. Then in the late 1980s and 1990s research showed us that children come to us with preconceived ideas of how the world works, and as idiosyncratic as these ideas sometimes are, they become a paradigm changer for education. A new question arose that had never appeared before: How do we deal with children who already have knowledge that is resistant to change but is not consistent with the current scientific knowledge (Driver, Guesne, and Tiberghien 1985)? How do we go beyond inquiry to inquiry for conceptual change?

New paradigms led us, as authors of this book, to reflect on our own teaching practices and how our own "nature of science teaching" changed. Was this serendipity? Was it conceptual change? Was it the fact that both of us were in a dark room of dissatisfaction with our then teaching style? Perhaps the answer is "yes" to all three. Our old styles no longer seemed to hold any relevance. We found ourselves, as did the scientists of the past, accepting new ways of looking at teaching as we became dissatisfied with the status quo. Our new "nature of science teaching" mirrored the common attributes of science as seen in the following table:

| Common Attributes of Science | Common Attributes of Science Teaching |
|---|---|
| Science is more than a collection of facts. | Science teaching is more than lecturing or doing hands-on activities. |
| Science knowledge is made up of current understandings of our natural systems and is subject to change. | Knowledge of science teaching is based on current understanding of effective teaching and learning and is subject to change as new research becomes available. |
| The body of scientific knowledge is constantly being revised, extended, and refined. | The body of knowledge about science teaching and learning is constantly being considered, expanded, and refined. |

## Author Vignette

I remember vividly trying, for the first few years of teaching middle school back in the 1960s, to be just like my cooperating teacher (the teacher who had supervised my student teaching). Then, one day while standing in front of a ninth-grade biology class lecturing about blood types, I had an epiphany. "Wait a minute," I thought, "Why am I explaining this when they could be figuring it all out for themselves?" I stopped talking, and set them to working on their own, and behold! They *did* figure it out for themselves. Was a paradigm shift or revolution there? Whatever you might call it, my teaching was never the same.

—Dick Konicek-Moran

## Author Vignette

In the early 1990s, I had a completely different epiphany from Dick's. I was teaching middle school and was considered to be a very good science teacher, doing a lot of inquiry and exciting things in my classroom. I happened to read a journal article in the *Phi Delta Kappan* on "Teaching for Conceptual Change" (co-written by Dick, as a matter of fact), and it spoke to me in a way that changed the way I viewed inquiry. My students were doing a lot of "hands-on science" and investigating their own questions, but I didn't start by finding out what ideas they already had in their heads (uncovering their preconceptions). I asked my eighth graders a question similar to the one in the *Kappan* article: What would happen to the temperature on a thermometer if it was placed inside a mitten? Surprising for me at that time, my students had the same misconceptions and theories as the fifth graders in the article I read. They, too, thought it would get "wicked hot" in there. I continued teaching the lesson in much the same way as Deb O'Brien, the fifth-grade teacher in the article. I let the students develop their own ideas, test them, and experience the cognitive

> dissonance that occurs when observations do not match strongly held predictions. When they were stuck, I guided them toward considering and testing an alternative idea. It was the best science lesson I ever taught (and probably one of the most memorable for my students). From that point on I transitioned from "hands-on inquiry" to "inquiry for conceptual change." My teaching style underwent a revolution and completely changed, as indeed did my career as I began to develop the formative assessment probes that today are changing the way science educators approach science topics and their teaching by starting with their students' ideas.
>
> —Page Keeley

The epiphanies we, the authors, had about teaching are just what we all want for our students—epiphanies that change the way they think about science. We are asking students to take up residence in a different world, a world steeped in science and technology. How do we get them to travel to this strange and uncertain, but rewarding, world? It would be unprofessional to leave it to chance and serendipity. History tells us serendipity occurs too rarely. Do we need to create dissatisfaction in their current theories so that they are open to more current scientific theories? Do we help them use conceptual models and out-of-the-box thinking to make the jump to new theories? After all, in 1865, Friedrich Kukele solved the puzzle of the structure of benzene by daydreaming of a snake with its tail in its mouth and came up with a circle as a model of the benzene ring, although it was not confirmed until 1929 through crystallography. Teaching students *about* the nature of science is only one part of preparing them for a scientific world. Having them *experience* the nature of science through their own conceptual change process is the other, which we will address in the next chapter.

In summary, developing conceptual understanding of the nature of science supports the learning of science. In the science classroom, students are using many of the practices of science described in the *NGSS*, such as asking questions, making observations as they conduct investigations, using models to predict and explain phenomena that are not directly observable, constructing explanations based on evidence, and engaging in argumentation with peers and the teacher to seek and solidify a common understanding. As students experience these practices in the classroom, they may take them for granted as being a routine part of their science instruction. Yet, these practices afford opportunities to point out to students that the aim of engaging in these practices is to establish well-accepted explanations that will help us understand observed phenomena and the behavior of objects, materials, and substances, rather than simply accumulate new knowledge.

## Questions for Personal Reflection or Group Discussion

1. Why is understanding the nature of science important to you as a consumer of science and scientific information? Why is it important to you as a teacher?
2. Which of the three models of the nature of science do you find most appealing and why?
3. How do you feel about using the term *the scientific method*? What terminology do you use to refer to the way scientists investigate?
4. Why is it important to acknowledge that scientists don't know everything? How can you model this in your classroom?
5. Jane Goodall is an example of a scientist who gathered data with an eye on using them for future research. How can you encourage this with your students?
6. If asking good questions is an important part of doing science, how can we make sure that our students realize that it is important?
7. Was there a time when you experienced a major shift in the way you teach science? What lead to this "revolution" for you?
8. How is teaching about the nature of science different from experiencing the nature of science? How do children's changing ideas mirror what happens when scientists modify their theories or propose new ones when new information is available? How does the process of discarding a misconception mirror what happens when scientists discard a long-held theory?
9. Choose one "golden line" from this chapter (a sentence that really speaks to or resonates with you). Write this on a sentence strip and share it with others. Explain why you chose it.
10. What was the biggest "takeaway" from this chapter for you? What will you do or think about differently after reading this chapter?

## Extending Your Learning With NSTA Resources

1. Read and discuss the chapter "Doing Science" in *What Are They Thinking? Promoting Elementary Learning Through Formative Assessment* (Keeley 2014).
2. *Uncovering Student Ideas in Science, Volume 3* (Keeley, Eberle, and Dorsey 2008) includes three nature of science probes: "Doing Science," "Is It a Theory?" and "What Is a Hypothesis?" Read and discuss the teacher notes that accompany each of these probes.
3. Examine the *NGSS* for nature of science connections. Browse the standards at *www.nextgenscience.org/next-generation-science-standards* and explore *NGSS*-related resources at *http://ngss.nsta.org*.

4. Read and discuss this article about how the nature of science can be addressed in elementary classrooms: Lederman, J., and N. Lederman. 2005. Nature of science is… *Science and Children* 43 (2): 53–54.
5. Read and discuss this article about how word choice (e.g., *the* scientific method) leads to misconceptions about the nature of science: Schwartz, R. 2007. What's in a word? How word choice can develop (mis)conceptions about the nature of science. *Science Scope* 31 (2): 42–47.
6. Examine and discuss the *Atlas of Science Literacy* maps for the nature of science. View them here: *http://strandmaps.nsdl.org/?chapter=SMS-CHP-0857.95*
7. View the NSTA web seminar The Nature of Science in *NGSS: http://learningcenter.nsta.org/products/symposia_seminars/NGSS/webseminar40.aspx.* The webinar is presented by Dr. Norman Lederman, one of the leading researchers on the nature of science.

# Chapter 4

## How Does the Nature of Children's Thinking Relate to Teaching for Conceptual Understanding?

During the last few decades, research on learning combined with the wisdom of practitioners have given us important new ideas to consider about deepening students' understanding of scientific principles. This chapter will look at these ideas, the past and current research behind them, and implications for the classroom when teaching for conceptual understanding.

### Education Research: How It's Done

We would like to take a moment to explain the ways that researchers examine teaching practices, since a lot of the ideas in this book are based on past and current research. Although education research *is* a scientific endeavor, it's not quite the same as an experiment done in a laboratory because, well, classrooms aren't carefully controlled laboratories! In many ways, they are more complex. Children and teachers come from different backgrounds; there are often differences in gender, economic status, motivation, and learning styles plus many more variables in location, time, and social environment. These and possibly more may be controlled or monitored by statistical analysis, depending upon the type of research.

Usually, the researchers define the problem they are trying to understand. Often, it involves a question such as, Does a particular teaching technique help students to understand certain concepts better than another technique? Terms need to be defined. What does *understand* mean? How will it be tested: by interview, paper-and-pencil test, or by task? The researcher then conducts a review of the literature surrounding this problem to find out what is already known in this area; he or she analyzes that research for validity and applicability. If various people are involved in collecting data, there has to be a common protocol so that each person is doing data gathering in the exact same way. Then the data are analyzed, depending upon the type of research being done. Some require statistics and others a rubric (a sort of rating sheet agreed on by researchers) for comparing data from one set of data to the next. For example, a case study (an in-depth scrutiny of some kind) may require an analysis of how a class or individual responds to a change in instruction over time. If groups of students are being compared, the researcher tries to make sure both groups are as closely matched as possible in every respect. Or a researcher may be interested in a dialogue on problem solving. The recorded interview will be analyzed to find out if the student changes his or her thinking when asked certain types of questions. The analysis will depend strongly on a set of pre-established expectations for different kinds of responses.

There may be *quantitative* research conducted, such as comparing two different types of instruction and the kinds of learning that happened in each. Analysis is usually statistical in nature. Examples may include data from surveys or test scores, or correlations among groups or individuals and the variables that are being studied. Another type of research is classified as *qualitative* and is used on variables that are difficult to measure; these include case studies, interviews, conversations, and other situations where the data are descriptive. Data from this research are often analyzed using a standard rubric. This would include anthropological techniques such as the researcher being a *participant observer*, where she or he becomes part of the group being studied and observes what happens in certain situations. For example, an observer may be interested in how and why a group within a classroom behaves in a certain way and what kinds of phenomena happen before the behavior occurs.

Teachers often involve themselves in *action research* in their own classrooms. Of course, this goes on all of the time as teachers try out different techniques and see if there are differences in results. Teachers make hundreds of decisions every day about why and how they approach a class. Action research applies more deliberate attention to prior research on a particular question, and to attending to variables so the results can be published or passed on to colleagues in some manner.

In this book, you will find many citations from researchers who have had their research reviewed by peers. In some journals, only 30% of the research submitted to the journal will be published. Each piece of research is sent to peer reviewers for analysis and either suggestions for changes are made or the research is rejected out-of-hand because of poor quality. Some research is collected in books by editors, and in other cases research is either presented orally at conferences or is published in journals, such as the *Journal of Research in Science Teaching, Science Education, School Science Review, Journal of Teacher Education,* and the *Review of Educational Research,* to name a few. Organizations such as the American Educational Research Association (AERA) and the National Association for Research in Science Teaching (NARST) meet yearly, and researchers share their findings with their peers. Each has its own journal on education research in science teaching. When you see citations in this book about research done in science education, you will know that the research has undergone strict review.

## What Are Children Capable of Learning in Science?

Much has changed in the past decade in our knowledge about children's capabilities in learning science. Research has shown that children are able to do many things earlier than previously thought (Michaels, Shouse, and Schweingruber 2008). Students come to K–3 classrooms with abilities in thought processing and reasoning that were previously viewed to be limited to upper-elementary and middle school children. In addition, we live in an information age where TV, the internet, and access to recent work of scientists are available to everyone. Educational TV programs such as *Nova* tell us about the current explorations and inventions of scientists and engineers. Thus, we must assume that children enter the classroom with well-developed theories about their world.

For instance, education researcher Kathleen Metz from the University of California, Berkeley, has shown that young children previously thought to be mired in a "pre-operational world" are capable of much more complex thinking (1995, 1997, 2004). (*Note:* The *pre-operational* stage of a child's life is a label from an extensive theory of child development by psychologist Jean Piaget, whose theories influenced a generation of educators from the 1950s to the 1980s. Piaget believed that, among other things, children at this stage were unable to understand logic or manipulate information mentally.) Metz found that some children come with fully fleshed-out theories about their world, but that many of these models are incomplete. In an article published in the *Review of Educational Research* in 1997, she reviewed the Piagetian research and found that she had questions about whether children were capable of theorizing at a young age even though they had limited knowledge in specific subject domains. She interviewed 48 preschoolers, evenly distributed for age and gender, and gave them nine problems to solve using different weights and a pan balance. To her surprise, the children showed both an understanding of the concept of *weight* and the ability to problem solve. This information led to our present belief, reflected in *A Framework for K–12 Science Education* (NRC 2012) and *Next Generation Science Standards* (NGSS Lead States 2013) that children can begin to use theories and problem-solving techniques at a much earlier age than previously believed.

We are continually learning more about what children can understand and do. What does this mean in terms of teaching for conceptual understanding? To consider this question, let's first examine some of the major contributors to the literature about thinking for conceptual change.

## Influential Models of Conceptual Development

### *Piaget*

Jean Piaget (1896–1980), a Swiss epistemologist, probably had more influence than any academician for almost 30 years. Although he wrote primarily about his observations about a very few children and was concerned mainly with development and not learning, many of his theories were misappropriated to make us believe that all children were incapable of thinking outside of what he considered four stages of development that were more or less age dependent. Before Piaget's work it was thought that children were just incompetent thinkers compared to adults. His work showed that children merely think in ways different than adults. He identified four stages of development: *sensory-motor*, which spanned the ages of birth to 2 years old; the *pre-operational* stage, where children were totally egocentric lasted until the age of 7; the *concrete operational* stage, where children could operate on physical objects and not on mental models; and the ultimate *formal operational* stage, where children were able to manipulate mental variables and to operate on materials in a logical manner. This stage began at about age 11.

Although Piaget is remembered more for his developmental stages, he also developed the theories of schemas, accommodation, equilibration, and assimilation that showed that schemas (the units of intelligent thought) were put into play as a child met a situation that was not understood. This caused *disequilibrium*, and the child had to *accommodate* the schemas by changing them to meet the new situation. This new accommodation resulted in *assimilation* of the new schema, thereby achieving *equilibrium*. For instance, a child might recognize *dog*, and therefore have a concept (schema) of a dog. This child meets a cat and calls it a dog. However, the behavior of the cat, and possibly adult intervention, causes *disequilibrium*. Thus, the child has to accommodate his dog schema to the new animal and thus reach *equilibrium* by *assimilating* the cat as a different animal. Piaget called this process *equilibration*.

In later years, Piaget (1973) considered himself a constructivist, believing that knowledge was created by the individual and often referred to his early work as his adolescent musings. Piaget's methods, however, of closely listening to the subject of research were seminal in education research.

### *Constructivism*

Constructivism is a central tenet to our (the authors') beliefs as educators. The fundamental principle of constructivism is that people construct their own meanings to make sense of their experiences or things that are told to them. The constructed meaning depends on a person's existing knowledge, which is affected by their experiences and the things they read or see and hear in the media as well as conversations with others. Since everyone has different experiences, has read or encountered different ideas in books and the media, and has talked with different people, it is likely that people will have different meanings for concepts, although there are similarities in ways of thinking about a concept (which is also why many of us have different meanings for the concept of *constructivism*). Some of the better-known education theorists who have espoused constructivism in their work are John Dewey, Maria Montessori, George Kelly (originator of *Personal Construct Psychology*), and David Kolb (of the *Learning Style Inventory*). One particular researcher, a colleague of Dick's at the University of Massachusetts, Ernst Von Glasersfeld (1917–2010), was a great theoretician and philosopher who was instrumental in framing the educational tenets inherent in constructivism during the 1990s and early 21st century (von Glasersfeld 1989).

### Author Vignette

I will often begin a professional development session by showing a cartoon of a young girl pantomiming the concept of a flower for her classmates. As the young girl is thinking and acting out "flower," the thought bubbles of the students do not show that they are thinking "flower." Instead, there are thoughts of different animals, forms of transportation, things blowing in the wind, and other objects moving. When I ask the teachers to discuss how the cartoon is like their classroom, they often bring up how students are full of ideas and that they have different experiences and background knowledge that they use to make sense of what the teacher is teaching. They further explain the cartoon by pointing out that the students' ideas are often quite different from what the teacher intended, but they make sense to the student. The cartoon always provides a nice segue into learning about how students construct their own understanding of concepts or phenomena by connecting them to something they already know or think they know.

—Page Keeley

## Current Models of Cognitive Development and Conceptual Change

Several different theories about knowledge acquisition and conceptual change are found in the education literature. These theories explain how concepts are created in individuals and help us understand what it takes for students to change their thinking. Five of these theories or models are summarized below.

### *The Theory-Theory Model*

We mentioned researcher Susan Carey in Chapter 2. Dr. Carey is a professor of psychology at Harvard University and is considered an expert in language acquisition and human cognition. She offers the "theory-theory" model in cognitive development to explain concept origins and conceptual change. Theory-theory proponents hypothesize that children are constantly creating theories to explain their world. Carey's model stresses that naive knowledge has the power to explain concepts across broad domains, such as the example in Chapter 1 of students who used the idea of population density to explain density at the atomic or molecular level. Thus children's preconceptions are often very hard to dislodge (Carey 2009).

### Phenomenological Primitives (P-prims)

In their article in the *Journal of Learning Sciences*, a group of researchers of constructivist learning led by John P. Smith III (Smith III, diSessa, and Roschelle 1993) from Michigan State University suggest that preconceptions are necessary, and through a process of knowledge refinement, can lead to attaining current scientific understanding. In this work, Andrea diSessa, from the University of California, Berkeley, theorizes that our everyday experience with physical phenomena leads us to collect commonsense intuitive ideas that we use to explain these encounters. diSessa calls these "phenomenological primitives" or "p-prims." They exist below the level of consciousness and can be persistent and resistant to modification. He believes that learning involves prioritizing and modifying these p-prims into ideas that reflect a way of thinking about phenomena in a realistic manner. He suggests that we abandon those that are not useful and bring to the forefront those that are. Reorganizing these p-prims into productive schemas helps the student become more expert in explaining the physical world. Students modify their thinking gradually so that ideas become "increasingly adequate."

These researchers believe that primitive ideas make up the theories of untutored children and adults and should therefore be the target of those who would change them. This is accomplished by discussion, not confrontation, since from a constructivist point of view all new knowledge is built upon prior knowledge. Attempts at replacement of one concept by another are not advocated nor deemed possible. Rather, use of analogies and sound questioning techniques are seen to be more productive (we'll go into this in more depth later in this chapter).

diSessa believes in what he calls *smaller grain axioms*, from which we build and modify concepts. A common example would be that of "proximity," since the novice realizes from experience that the closer one gets to a source of heat energy, the warmer one feels. This would lead novices to form a conceptual scheme that would explain the seasons as being caused by the Earth being closer to the Sun in summer and further away in winter. diSessa would not call this a misconception but rather a *miscued p-prim* that was brought to consciousness and applied in an inappropriate manner.

### Conceptual "Facets"

Along this same line of thinking, Jim Minstrell, a teacher-researcher in physics education, used a similar notion to p-prims in his work begun during the 1980s and continuing today, called "facets" (1992). Minstrell began studying the learning and teaching of his own students and those of his teacher colleagues. He realized that there were strengths as well as problematic thinking in his students' understandings, which prompted Jim and colleagues to coin the term *facets of student thinking* to replace the pejorative term *misconceptions*. Minstrell's facets are a unit of thought, a piece of knowledge, or a strategy. A typical facet might be "passive objects do not exert forces." Identifying these facets and trying to understand their origin and meaning are helpful in developing ways to help students to modify their conceptions to fit those held by scientists.

### *Causal Models*

In their paper published in 2005, David Perkins and Tina Grotzer suggest that helping students understand what kind of causal models they are using helps them toward deeper understanding of science concepts. They identify four different levels of causal models and show how students can view phenomena in different ways. Their premise is that students are familiar with only the simplest type of causal model—for example, "*A* affects *B* affects *C*." More complex causal models such as those used to understand predator-prey relationships in an ecosystem or the analysis of how an electrical circuit works require more complex analysis of causal models (Perkins and Grotzer 2005). The first example of a causal model that comes to mind is a cause-and-effect model, such as "Those who exercise regularly will live longer than those who do not." Statistical evidence can be gathered to pit exercise against longevity.

Another example might be in an ecosystem where the addition of a predator changes the behavior of prey and thus changes the ecosystem. A perfect true-life example is that of the introduction of wolves into Yellowstone National Park. The wolves prey upon the elk that eat aspen saplings in the forest. Observers have noticed an increase in the aspen forests and have theorized that the presence of the predators has caused a change in feeding behavior of the elk; they either frequent different forests or eat less because they are on the lookout for the wolves. Thus, the aspen forests grow due to less foraging on the part of the elk. "*A*" (the wolves) affects "*B*" (the elk) affects "*C*" the aspen. Also noticed is that nearby rivers, "*D*" changed course due to the effect of the changes in aspen growth. All of this has been designated as a "trophic cascade." Research is still being done on this model of interactions to see if it is consistent.

### *Conceptual Ecology*

It may be helpful to think of children's ideas in yet another way. In 1972, British philosopher and logician Stephen Toulmin suggested that concepts do not exist in isolation but rather in what he termed a *conceptual ecology*. We cannot hope for conceptual understanding by attacking a single concept if it belongs to a group of interconnected concepts or schemas. If we use an ecological metaphor, a change in one aspect of the ecosystem will trigger a change in all parts of the ecosystem, and the changes may not be for the best or change in the direction we would wish. Kathleen Roth, from Michigan State University (and now at BSCS), took this idea and expanded on it. She believes that change in the ecosystem may itself trigger a response in the student to modify further his or her conceptual ecology and hold onto the original idea even more strongly or, as Roth suggests, change the entire "ecosystem," leaving the student completely confused and at sea. She also suggests that a student with a strong central schema may choose to ignore new information because it would destroy the ecology of the preconception. It's easier to do this rather than completely reorganize the ecosystem (1990).

## Sorting Through Conceptual Change Models

You may be thinking, "Oh no! There are so many different views on how children learn. How do we choose which one will best help us understand our students?" Welcome to science, where conflicting theories are often the norm!

We can find some basic commonalities in the theories of researchers summarized above. They assert (1) learners form their knowledge from the experiences they have with their environment. They also believe (2) naive units of thoughts (facets, p-prims or pre-concepts) influence learning. And they all suggest (3) we must allow more time to work on students' conceptual change since students' ideas about concepts are so resistant. Thinking of conceptual ecologies helps you understand why children's beliefs are so difficult to change. We also know that students have similar patterns in their preconceptions and that each student is not entirely unique in their ideas (Shapiro 1994), which brings us to the last area of research on conceptual change.

### *Alternative Conception Research*

Known by a variety of names: "misconceptions," "pre-instructional ideas," and "naive constructions" to name a few, alternative conception research is generally taken on by practitioners within their own domains. The point is to try to establish if there are common patterns of preconceptions within numbers of children that teachers may notice. John Clement and Jim Minstrell, working in physics education, are well-known researchers in this field, and you will find references to alternative conceptions found in other scientific fields or domains throughout this book as well. The work of Rosalind Driver and her colleagues in the Children's Learning in Science (CLIS) project at Leeds University in the United Kingdom is also a landmark reference for exploring children's ideas in science.

## How to Effect Conceptual Change

So, there are many theories about how we develop and hold onto ideas in science. As teachers, we need to help students change their naive conceptions (or pre-prims or facets) into concepts that more closely mirror current scientific thinking.

Early education researchers, lead by George Posner of Cornell University (1982), suggested that conceptual change was dependent on several aspects. Change would occur if

1. the student was dissatisfied with the current concept;
2. a new concept appeared to be believeable and reasonable;
3. the new concept was understandable; and
4. the new concept was fruitful.

### *Conceptual Change Versus Conceptual Exchange*

However, Peter Hewson, from the University of Wisconsin, suggested that there is a difference between two terms that separate the kinds of change we can expect: *conceptual change* and *conceptual exchange*. In the former case, the concept that was changed still remains and in the latter, it is replaced and is gone. He goes on to say the following:

> *I ... see conceptual change as primarily a way of thinking about learning, i.e., it is something that a learner does as an intentional act, rather than something done by a teacher. There is, of course, much that a teacher can do to facilitate a student's learning, without any need to regard this as a mechanistic, causal process. Finally, it seems to me that the knowledge a learner gains only has validity in terms of, and is thus relative to, his or her conceptual ecology. Since a learner's conceptual ecology is a product of all the experiences and social interactions he or she has had, it will have many elements in common with those of other people. (Hewson 1992, p. 9)*

This means that a curriculum should include not only particular theories and attendant phenomena but also the basis for their acceptance. If we can't justify curriculum content to students, we shouldn't teach it. In other words, we need to recognize that "alternative" is not a synonym for "inadequate" or "unacceptable." The purpose of teaching for conceptual change in science is not to force students to surrender their alternative concepts to the teacher's or scientist's conceptions but, rather, to help students both form the habit of challenging one idea with another and develop appropriate strategies for having alternative conceptions compete with one another for acceptance.

Thus, Hewson believes, as we do, that it is the students who must make the change based on their own analysis of their alternative conception versus evidence available to them regarding a new concept. Hewson's point about "developing appropriate strategies" is explicit in the emphasis on scientific practices in *A Framework for K–12 Science Education* and *NGSS*. In Chapter 7, we will look at instructional models that suggest that the change of concepts is gradual and done in small increments, rather than making a large change all at once. As an example of the difficulty of effecting conceptual change, let's take a look at Jon, a middle school student featured in a video from *Minds of Our Own*, the "Lessons From Thin Air" episode (Harvard-Smithsonian Center for Astrophysics 1997). Jon, despite receiving what could be considered by common standards excellent instruction, held on to his primitive beliefs about photosynthesis and plant biology. Jon believed food for a plant (including soil and water) came via the root system. When questioned by a very capable interviewer, Jon was determined that air had no substance, even when presented with a piece of solid carbon dioxide in the form of "dry ice" and seeing a difference in the weight of a balloon empty and then blown up. His comment was, "Sometimes air has mass and sometimes it doesn't." His beliefs were probably based on a lack of understanding of the particulate nature of matter and that air had substance. How could carbon dioxide, which he perceived as a massless substance, provide enough carbon to make up the mass of a tree? Years later in an interview (privately seen) Jon appeared with his chemistry book in hand, with a new ecology of concepts and talked with great humor about his prior interview. Somehow, he had effected conceptual *exchange,* and he now believed that air had mass and matter was made from particles that he could now "see" or at least imagine. The interview with Jon as well as several of the other videos in the *A Private Universe* series can be viewed at *http://learner.org*.

## Author Vignette

Somewhere about 1993, Harvard-Smithsonian Center for Astrophysics media group decided to promote a means for professional development via an interactive series of programs called *A Private Universe*. Matt Schneps and Philip Sadler, via Nancy Finklestein, invited me to be a content guide for two of the nine programs. The series was sent out on live television to teachers and educators all over the world. The participants had an opportunity to call in to us since we were televising live. Being a content guide included being a part of planning the programs and using the vast library of videotapes of interviews with students on their ideas about science concepts. It was a wonderful learning experience to me to be part of this new technological exploration of professional development. It basically amounted to two hours per program of working with educators in an open-ended arena and responding off-the-cuff to comments and questions from educators who were learning right along with us. Unknown to me, Page was one of the participants and it was years later that I learned that she witnessed my television debut.

Over the years, I also had the opportunity to consult with the Media Group many times as they presented more professional development programs, such as *Looking at Learning* and *Looking at Learning Again*. It was always a pleasurable and enlightening experience to work with these learned people who worked on the cutting edge, helping teachers to elicit the thoughts and concepts of students and help them to grow into educated and competent science teachers.

—Dick Konicek-Moran

## Techniques for Teaching for Conceptual Change

The challenge to teachers and administrators at all levels is to meld our new knowledge about how children learn into a comprehensive and integrated form of curriculum and instruction. Teachers and administrators will need "models of classroom instruction that provide opportunities for interaction in the classroom, where students carry out investigations and talk and write about their observations of phenomena, their emerging understanding of science ideas, and ways to test them" NRC (2007).

Let's look now at some teaching techniques used currently and what some education researchers have to say about them.

### *Listening and Probing*

Dr. Bonnie Shapiro, whom we introduced in Chapter 1, demonstrates that teachers are often fooled into believing that everyone in a class is on the same page, when they are not. Shapiro interviewed a sample of six fifth-grade students whose class was studying light. They all had created their own mental models about how the eye works (mostly that seeing originates in the eye) and were unwilling to modify them even during and after instruction. One student even called the teacher's explanation "silly" and "unbelievable." One important point is that they did not need to change their ideas in order to succeed in school. When asked how they knew the answer the teacher wanted, they responded, "Mr. Ryan (fictitious name) always wrote the correct answer on the overhead projector." Many students believe that going to school is their job and that their job is to complete assignments, receive a grade, and move on. They were able to complete the tasks assigned to them without having the basic concept in hand. They knew how to play the game of "school."

The teacher was unaware of the situation and continued to use the familiar activities such as the color wheel, refraction of a pencil image in a glass of water, and so on that we in science education refer to as *critical experiences*, which he believed would bring the students to an understanding of the concept of *light*. Since the students were merely going through the motions of doing the activities and filling out the worksheets, they had no opportunity to develop a network of ideas that could accept the results of the critical activities. Eleanor Duckworth, in her book *Tell Me More* (2001), says, "'Critical' experiences are seen to be critical only to a person who has already developed a network of ideas that give the experience meaning. The same is true of any idea that does not connect with the ideas a learner habitually uses" (p. 185).

During class discussions, only those students who understood the canonical ideas concerning light voiced their opinions and engaged in conversation with the teacher. The other students were loath to say anything contrary to the teacher. He was therefore unable to "listen" to the many alternative concepts that were lurking in student minds and undermining his attempts to teach for conceptual understanding. Had he known the status of his students' thinking, he might have designed lessons that would inform him of his students' ideas.

Building student trust and engaging the class in group discussions is the principle way a teacher can "hear" what is going on in the students' minds. Mr. Ryan might have begun with a probing question that would have informed him of his students' preconceptions so he would have been able to plan how to approach his instruction. A great many of these probes have been developed in Page Keeley's *Uncovering Student Ideas in Science* series.

Two of the most important new watchwords for teaching for conceptual understanding will be to *listen* and *probe*. Children's alternative conceptions are very important and teachers need to try to understand them, including how the child came to construct them. As we have said before, alternative conceptions, no matter how naive or seemingly incorrect, are the foundations for building new and more complete conceptions. They provide us with a place to start teaching. Listening to children can reap many benefits, among them the realization that teachers are more than just teachers; they are teacher-researchers, engaging in informal research about students' learning and thinking.

Because of this, one of the most important processes embraced by educators during this decade is formative assessment. As we stated before, we find few problems with the idea of testing for achievement or the idea of accountability (summative assessment), but it is evident that unless teachers know where their students are in their current conceptual development, they cannot help their students make changes in their thinking during the instructional process, before the students' knowledge is measured on a test or other summative assessments. This means we have to give them opportunities to tell us about their thinking: how they use vocabulary with understanding and access concepts—the "network of ideas" to which Eleanor Duckworth refers. We will discuss more about formative assessment in Chapter 9.

Students need to relate the material they are studying to everyday events. Students have trouble applying material learned in school-centered science lessons to their real lives. As authors of other series, we have deliberately tried to do this in our work. Dick has written a series of books, *Everyday Science Mysteries*, so students can connect their school learning with their real lives. Page's *Uncovering Student Ideas in Science* series sets most assessment probes in an everyday familiar context so children can relate their thinking to their real-life experiences.

### Seeking Common Understanding by Asking Questions

Often when children and adults talk to each other, there is a problem of incommensurability: two people in a conversation not speaking the same "language." In education, not only are teacher and student often using different meanings of the same language, but they are also operating in different paradigms or rules about how the world is seen and studied. Children focus on different things and on different questions than do adults.

George Hein tells a story of a group of students presented with an inclined plane and a set of empty cans of various sizes. What the teacher had planned was to show that the cans traveled at the same rate down the ramp despite their size and mass. Children were instructed to roll the cans down the ramp in pairs and record the results. Naturally the students began to roll the cans down the ramp to see if there was a can that rolled faster than the others. When asked what they found out, they stated that they had rolled the cans 10 times and that one can went faster than the others. When questioned further about their evidence, they stated that a particular can had "won" the race twice and, besides that, there were eight "ties." The data, as far as they were concerned, pointed to one can as the "winner." The children saw the exercise as a race, and in a race there must be a winner. Therefore, the entire point of the lesson—that most of the cans traveled at the same pace no matter their size—was lost to them. No wonder Hein entitled his article "Children's Science Is Another Culture" (Hein 1968).

Thus, talking about what happened resulted in exposing the incommensurability between the teacher and the students. They noticed different things and focused on different results. So much of what children think about phenomena depends on what they notice when they observe. For teachers, it is important to know what it is students notice and on what they focus their attention.

## Author Vignette

After reading the article by Hein, I decided to try this out with a group of third graders. As I interacted with the children as they rolled the cans down the ramp, I was struck with how deeply they felt pressed to declare that one can "won" the race and were reluctant to even consider anything that was a tie. They had already made up their collective minds that ties were not acceptable in this "game." After discussing the results with them, I was informed that if there was not going to be a winner, there was little use in doing the activity. When I viewed this as research, I found out a great deal. When I viewed this as a lesson, I realized that I came up lacking and might have used an advanced organizer, perhaps a recording sheet that pointed out the expectations of "ties." Or I might have talked to them more about the purpose of the activity. As it turned out, our discussion after the activity allowed the children to discuss their results and realize that there was importance in the ties. I walked away with the feeling that there had been some conceptual change, but probably not much. As I ruminated on the events later that evening, I realized, as a teacher, how easy it was to delude myself into thinking that I had made a difference. I would have needed to go back later and try another similar activity and check the results.

—Dick Konicek-Moran

Cognitive psychologist Susan Carey believes that in learning that is truly conceptual change, incommensurability is to be expected. It is the teacher's task to overcome this. She does this by trying as hard as possible to understand the child's point of view; trying to get, figuratively, inside the child's mind and understand what the child is saying and thinking. Eleanor Duckworth has experimented with listening and questioning skills for many years. She says, "The more I was interested in their thoughts, the more they were interested in their thoughts and the more they got interested in pursuing an investigation" (Duckworth 2012).

Duckworth is famous for doing English translations for Jean Piaget whenever he spoke in his native language, French, to an English-speaking audience. She is also famous for her teaching strategies and books that depict ways to understand children's thinking by engaging them in meaningful dialogue. She suggests that we use such questions as, "What do you mean? How did you do that? Why do you say that? How does that fit in with what he or she just said? Could you give me an example? How did you figure that?" (Duckworth 1987, 2001). She goes on to explain that these types of questions are a way for the teacher to understand what the student understands by engaging the other's thoughts and helping them to consider other options. This is doubly true for children raised in different cultures than that of the adult.

In research done in the 1990s by John Leach, Bonnie Shapiro, and Dick Konicek-Moran (1992), hundreds of students were interviewed about their ideas regarding natural decay and decomposition. The following vignette describes Dick Konicek-Moran's experience interviewing one of these children:

## Author Vignette

In 1991 and 1992, I had the opportunity to work with Rosalind Driver's group at the University of Leeds in England. When I returned to the United States, my graduate students and I formed a group called the "Dead Fruit Society." We interviewed hundreds of students and each student was asked to draw or describe what would happen to an apple that was allowed to remain on the ground for a full year. We were astonished to find that there was little difference between the answers of six-year-olds and those who had recently taken a biology course in high school. No one mentioned decomposers, and none seemed aware that the apple would be decomposed into molecules that could be used by other plants. Later, another graduate student set herself up in the local mall and asked the same questions of adults with the same result.

As a sideline, during the interviews, a fifth-grade female student told me that her grandmother put fishmeal on plants to make them grow better. I asked her if she had asked her grandmother why this helped growth. She looked at me as though I were addled. "No child would question and elder about things like that," she answered indignantly. I learned two important things from that encounter: one, the girl probably did not know any details about the role of fertilizer and two, I would not expect the girl to question anything an adult told her.

—Dick Konicek-Moran

It is incumbent on the teacher to allow students to express their knowledge about the subject at hand—which brings us to a discussion about talking and language.

## Talking and Language

Lev Vygotsky, a Russian psychologist famous for his pioneering work on learning and child development, theorized that language and thought were interconnected and that listening to a child thinking out loud was a window into that child's inner world. Vygotsky believed that children are initially guided by social speech (spoken by others), which then led to egocentric speech (outer speech for oneself, which somehow mirrors the social speech) and finally to an inner speech that belongs entirely to the child his- or herself. It is the inner speech that leads the child to engage in activities guided by his or her own thinking. Vygotsky believed that social interaction shapes intellectual development:

> *The greatest change in child development occurs when ... socialized speech, previously addressed to an adult, is turned to himself (sic), when, instead of appealing to the experimentalist with a plan for the solution of the problem, the child appeals to himself. In this latter case the speech, participating in the solution, from an inter-psychological category, now becomes an intra-psychological function. (Vygotsky and Luria 1994, p. 119)*

Vygotsky's early work is further supported by many recent studies that suggest that productive science talk has many benefits in supporting student learning in the classroom (Michaels, Shouse, and Schweingruber 2008). Productive science talk has been found to lead to students becoming engaged in the content of the lesson. It elicits some surprisingly sophisticated reasoning by students who may not be considered academically successful and leads to increased participation by students who may have been disengaged in learning science.

### *Dialogic Teaching*

Robin Alexander from Cambridge University in the United Kingdom suggests that dialogue that stresses purposeful questioning and the stringing together of ideas into meaningful thinking is necessary to what he has labeled *dialogic teaching*. Alexander says it is necessary for the following criteria to be present in the classroom to promote this kind of teaching:

1. Teachers use questions that suit the purpose of the lesson, some requiring a range of responses and inspiring divergent thinking, others requiring single word replies; and in that chaining of questions and answers, ideas are developed or changed.
2. Teachers encourage language and learning talk through activities that require children to respond in extended answers, and that teachers model language that is understandable yet challenging.
3. Teachers listen to the content of students' responses—challenging, probing, and extending their meanings; and offer constructive and formative feedback. In this process, both teacher and students participate in discussion that is alive with provisionality and even uncertainty. Children initiate debate where speaking turns are evenly distributed, and sometimes the teacher withdraws from the floor. Students are expected to address issues in this open forum in an intelligible manner and listen to each other's contributions. (Alexander 2008)

Another British researcher, Sylvia Wolfe, after observing classrooms and talking with primary teachers in the United Kingdom over three years, developed a similar list of important behaviors used by teachers to promote dialogic teaching.

Teachers are

- asking authentic questions;
- using deferring questions to check children's meanings;
- pausing to allow children time to think and interject and express ideas fully (see also Mary Budd Rowe [1986] on "wait time");
- adopting a low modality, using words such as *perhaps* and *might* as invitation to a range of possible actions;
- offering new content relevant to the theme unfolding;

- developing a line of argument by staying with one child through a sequence of connected questions;
- accepting responses without evaluating them;
- engineering opportunities for students to participate actively in the discourses; and
- building on children's interests. (2006, pp. 258–259)

### Argumentation

Argumentation is a form of talk that has important applications in science. There is a growing body of research that suggests that children learn more effectively and achieve more when they are actively involved in activity, through discussion, dialogue, and argumentation (Mercer and Littleton 2007). *A Framework for K–12 Science Education* and the *NGSS* include a strong and central role for the practices of explanation construction and argumentation in science classrooms (NRC 2012, NGSS Lead States 2013).

> *A prominent, if not central, feature of the language of scientific enquiry is debate and argumentation around competing theories, methodologies, and aims. Such language activities are central to doing and learning science. Thus, developing an understanding of science and appropriating the syntactic, semantic, and pragmatic components of its language require students to engage in practicing and using its discourse. (Duschl and Osborne 2002, p. 40)*

As we have pointed out, science is a social endeavor. Research has shown that there is a great difference between argumentation and confrontation (diSessa and Minstrell 2000). Winning is not the object of scientific argumentation but, instead, reaching consensus or coming up with an idea that makes sense to the community. This is contrary to what most people think of when they consider making an argument. Argumentation of course has some drawbacks. It may lead to stubbornness or even to the opposite, a conversational tone that ultimately leads nowhere. Therefore it is important to develop norms and ways of productive talk in the classroom. TERC's The Inquiry Project/Talk Science (*http://inquiryproject.terc.edu*) is an excellent resource for fostering and supporting productive classroom talk so that students can engage in the practice of evidence-based argumentation.

### Classroom Argumentation

Three NSTA Press books offer practical forms of scientific argumentation that can be adopted across disciplines and in all grade levels: *Scientific Argumentation in Biology: 30 Classroom Activities* (Sampson and Schleigh 2013), *Argument-Driven Inquiry in Biology* (Sampson et al. 2014), and *Argument-Driven Inquiry in Chemistry* (Sampson et al. 2015). The primary author, Dr. Victor Sampson, is widely published in the area of classroom argumentation. These books provide lessons in each of three categories: (1) generating an argument requiring students to make a claim about a research question where the teacher supplies data; (2) a model called "evaluating alternatives," where students are required to respond to a discrepant event, or two or more alternative explanations; and (3) writing activities that give the student an opportunity to write about and refute other points of view that exist in their world, state a claim, and defend it in writing. This also addresses the suggestions in the *Common Core State Standards,* in English Language Arts (NGAC and CCSSO 2010). Refutational writing originates from the Greeks and part of the *Progymnasmata* or elements of rhetoric (Kennedy 2003).

Evaluating alternatives can also be carried out in primary or middle schools by using the probes in *Uncovering Student Ideas in Science* series or the alternative ideas expressed in the stories in the *Everyday Science Mysteries* series.

## Author Vignette

I once witnessed a kindergarten class arguing about whether a predicted snowstorm that night would result in a snow day. The children held an impromptu vote, and since the vote fell in favor of a school-closing snowstorm, all left that afternoon believing that they would spend tomorrow happily sledding or playing in the snow. The next day, which to their consternation was spent in school, they had a discussion about predictions, evidence, and consensus.

—Dick Konicek-Moran

### *Attempting Consensus and Civility*

The goal, of course, in any group dialogue, and especially in science, is consensus. Scientists who engage in dialogue through conferences and responses to published papers do not vote to see who is right. If there is a change in a paradigm, it can be noticed in a shift in the kinds of research or in the topics that are chosen for research publications. Both of these are indicators of consensus (Kuhn 1996).

Much of the talk that goes on in classrooms is of a nature called a *triad*. The teacher poses a question, the student answers, and the teacher responds in some way, either to correct or to comment further. Eichinger and colleagues have found that argumentation in a classroom is more likely to be productive when students are encouraged to interact with each other rather than having their discussion filtered by a teacher. They also found that it is important for the teacher to help students focus on tolerance and to make sure that arguments are based on theory and evidence (Eichinger et al. 1991). Students need help in this and should not be expected to learn this by themselves (Osborne et al. 2001). Understanding civility is important when helping students argue productively. Students, even up to adulthood, are determined to "win" an argument. It is not easy to help students see the difference between arguing with an idea rather than arguing with a person.

Children can set their own standards for argumentative interaction and are capable of doing so. They must be aware that they are arguing about *evidence* rather than personal feelings. In fact "arguing" may well be a misleading term for classroom use. A better term might be *dialogue*. In fact, it might behoove a teacher in some classes, including those involving adults, to suggest a responder find something positive to say before finding fault with an argument put forth by another. Some teachers insist that the responder in a dialogue restate the previous speaker's ideas until that person agrees that she has been heard and interpreted correctly.

In 1912 Alfred Wegener suggested a theory of "continental drift" as an explanation for mountain building, volcano activity, and the shapes of the continents that seem to fit together. However, there were no data to support the idea that the seemingly solid and fixed earth could move. It took 50 years and a great deal of research before the idea of continental drift became plausible. The theory is still being debated and probably will be for years to come. Debate about theories is what science is all about. Debate about theories is what science learning in schools should be all about.

## Understanding the "Pain" of Changing Ideas

If we are honest with ourselves, no matter how objective we think we are, it is hard to change our minds when confronted with evidence that is contrary to our firmly held thinking patterns. We see how difficult, and even painful it can be to change our ideas. This depends on how entwined they are with other ideas, beliefs, wishes, and thoughts. The more entwined the idea, the more difficult to change, because changing ideas means giving up those that may have been important to us (Duckworth 2001, p. 185).

Teachers often feel committed to changing that "wrong" idea as quickly as possible by whatever means they have at their disposal. We are suggesting, instead, that this is the time to act as teacher-researchers and to create a safe space in the classroom to have dialogue and argumentation. We need to listen as carefully as possible and to question students to find out where the ideas originated and how deeply they are committed to them. We should make students see how interested we are in how they think and encourage them to consider their own thinking—to engage in metacognition. The conversation does not have to be one on one. Instead, we suggest that students talk to each other in small groups and then to the teacher out loud, bringing their thoughts to the front, so all students can hear. Delaney has found that students are very interested in each other's thinking and that extended conversations, even in a small group in the classroom, can push the entire class to think more deeply (in Duckworth 2001). By modeling an expectation of change in thinking, and the joy in actually doing it, we can mediate the discomfort they feel—the discomfort we *all* feel when we find our old ideas "wrong" or not useful anymore.

## Author Vignette

In my early years as a teacher of adults, I had my students involved in trying to explain why popcorn kernels rose and fell in a container of a carbonated liquid solution. After a long and animated discussion among the participants of the class, one adult came up with a reasonably plausible explanation with which many agreed. When one of the other students suggested that I give them the "real answer," she objected violently. "No," she said. "This is the first time in my life that I have had the chance, in a class, to come up with my own idea, and I would like to enjoy that for a while. I like it, and it works for me." The discussion continued and as it did, she observed some flaws in her model and modified it to fit her needs. I no longer remember if the final consensus was the "right" answer, but I shall never forget the student's comment and the way the group worked together in listening to each other's ideas and negotiating a plausible explanation while looking at their thinking, parsing it, and coming to an agreement.

—Dick Konicek-Moran

## Keeping the Long-Term Perspective of Learning

Although it is our goal to bring about conceptual change, it will not necessarily come within our purview as a classroom teacher for a semester or a year. Dick offers a story about a student who was enrolled in an introductory physical science class at the ninth-grade level.

## Author Vignette

The original introductory physical science class used balances that also had riders on the arm of the balance so that tenths of a unit of the standard mass could be calculated. During an activity of measuring the mass of a plastic vial, a fragment of an effervescent tablet, and the lid of the vial, the students were supposed to discover that when the ingredients were put together and a chemical reaction occurred in the vial, the mass of the entire closed system remained constant. After the activity was completed, I noticed that one of the students was very excited—actually jumping up and down. When I asked what the student found so exciting about the activity, the student replied, "It wasn't the activity. I think I finally understand decimals!" The use of the balance had triggered something in the boy's experiences that caused a conceptual change—not in understanding mass, but understanding something that he was supposed to have learned four years before in fifth grade! Up until that moment in ninth grade, he'd been fooling everyone, including himself, into believing that he had a working theory of decimal numbers.

—Dick Konicek-Moran

The point of the above vignette is that even though a teacher has done everything in his or her power to help a student change a conceptual view, it might finally happen in another place at another time. Prior teachers had helped build the intermediate conceptual changes, but the final spark that engaged the "lightbulb" was a long time coming. All those other teachers were not present when the "aha!" moment happened, but we must take comfort that prior experiences had something to do with the final result. We teachers all helped create Eleanor Duckworth's "network of ideas" in students.

## Questions for Personal Reflection or Group Discussion

1. We often hear the claim "research-based" in regard to teaching and learning. What does "research-based" mean to you?
2. Think of a problem or practice you would like to research in your classroom. How could you use action research to learn more about the problem or practice

by being a researcher in your classroom? How might you search the literature to find out more about the problem or practice?

3. This chapter talked about how we now know that young children are capable of more sophisticated scientific thinking than we previously thought was possible. How do you feel about raising standards for students? How does *A Framework for K–12 Science Education,* the *NGSS,* or new state standards reflect this shift in thinking about what young children can understand and do in science? Share some examples.
4. Of the various models for conceptual change, which model appeals to you and why?
5. Expand on the theory of conceptual ecology in a classroom. Try to think of examples of conceptual ecology in either your classes as a student, in your class as a teacher, or in your work with teachers as learners.
6. Teachers are often confronted with several theories about how to best teach science. How do you think you should choose among the various theories, given the limited time you have to teach science?
7. Select a unit that you teach. Imagine you discover a major misconception your students have as you begin to teach the unit (identify the misconception). What do you think is the best way to confront this misconception in your instruction?
8. Reflect on the difference between conceptual change and conceptual exchange. Can you think of an example from your own experience as a learner or a teacher?
9. Hearing and listening are not the same thing. What do you think is the difference? Do you spend more time hearing what your students say or listening to them? How can you be a better listener of children's ideas?
10. How would you set up an environment conducive to promoting productive argumentation in your classroom? What kinds of norms would you set with your class? How can you help students learn ways of "talking science?"
11. Choose one "golden line" from this chapter (a sentence that really speaks to or resonates with you). Write this on a sentence strip and share it with others. Explain why you chose it.
12. What was the biggest "takeaway" from this chapter for you? What will you do or think about differently as a result?

## Extending Your Learning With NSTA Resources

1. Read and discuss an article about teacher research into students' ideas: Keeley, P. 2011. Teachers as classroom researchers. *Science and Children* 49 (3): 24–26.

2. Read and discuss Chapter 3, "Foundational Knowledge and Conceptual Change," in Michaels, S., A. Shouse, and H. Schweingruber. 2008. *Ready, set, SCIENCE! Putting research to work in K–8 science classrooms.* Washington, DC: National Academies Press. This book is also available as a free pdf download at *http://nap.edu*.
3. Read and discuss two successive articles about constructivism and conceptual change: Colburn, A. 2007. The prepared practitioner: Constructivism and conceptual change, Part 1. *The Science Teacher* 74 (7): 10, and Colburn, A. 2007. The prepared practitioner: Constructivism and conceptual change, Part 2. *The Science Teacher* 74 (8): 14.
4. Read and discuss chapters on the science of learning in Bybee, R. 2002. *Learning science and the science of learning.* Arlington, VA: NSTA Press.
5. Read and discuss the guest editorial about conceptual shifts in the *NGSS*: Pratt, H. 2013. Conceptual shifts in the *Next Generation Science Standards*: Challenges and opportunities. *Science Scope* 37 (1): 6–11.
6. Read and discuss this article on how argumentation promotes learning in inquiry science: Sampson, V., and C. Hall. 2009. Inquiry, argumentation, and the phases of the moon: Helping students learn important concepts and processes. *Science Scope* 33 (9): 16–21.
7. Read and discuss chapters from this book on responsive teaching that focuses on shifting high school traditional teaching to listening for, identifying, and responding to students' ideas: Levin, D., D. Hammer, A. Elby, and J. Coffey. 2013. *Becoming a responsive science teacher.* Arlington, VA: NSTA Press.

# Chapter 5

## What Can We Learn About Teaching for Conceptual Understanding by Examining the History of Science Education?

This chapter will provide a very brief history of science education in the United States from the turn of the 20th century until now, the 21st century. We feel that this historical perspective is important in considering how to teach for conceptual understanding, especially now, when science teaching is on the cusp of change as a result of new considerations for teaching and learning with the *Next Generation Science Standards* (*NGSS*). While an entire book can be (and has been) written on the history of science education, we chose to highlight a few episodes and projects.

### John Dewey: The Prophet of Effective Science Education

To begin this chapter, let's go back more than a century ago. A great educational thinker named John Dewey had ideas about science education that still seem remarkably current. The following are quotes from an article he wrote for *Science*, the journal of the American Association for the Advancement of Science (AAAS) in 1910:

> *Science has been taught too much as an accumulation of ready-made material with which students are to be made familiar, not enough as a method of thinking, an attitude of mind, after the pattern of which mental habits are to be transformed. (p. 122)*

> *Science teaching has suffered because science has been so frequently presented just as so much ready-made knowledge, so much subject matter of fact and law, rather than as the effective method of inquiry into any subject matter. (p. 124)*

> *I do not mean that our schools should be expected to send forth their students equipped as judges of truth and falsity in specialized scientific matters. But that the great majority of those who leave school should have some idea of the kind of evidence required to substantiate given types of belief does not seem unreasonable. Nor is it absurd to expect that they should go forth with a lively interest in the ways in which knowledge is improved and a marked distaste for conclusions reached in disharmony with the methods of scientific inquiry. (p. 126)*

*When our schools truly become laboratories of knowledge-making, not mills fitted out with information-hoppers, there will no longer be need to discuss the place of science in education. (p. 127)*

Although Dewey's writings have been seen as a guiding light in education, his suggestions for science education have not always been heeded diligently over the years. We are hopeful that the *Next Generation Science Standards* will finally bring Dewey's century-old, timeless wisdom to the forefront of modern-day science education.

## The Golden Age of Science Education

Long ago in the 1960s (a distant memory, at least for one of your authors, who pioneered some of these programs in his earlier job as science coordinator for a school system), there were several science programs available to school districts, funded by the National Science Foundation (NSF) and designed to improve science instruction. They were developed during what we called, "the golden age of science education." After the Russians launched the satellite Sputnik in 1957, our nation became worried about being second best in space technology. So, the United States signed into law on September 2, 1958, the National Defense Education Act (NDEA) and poured millions of dollars into schools, curriculum programs, teacher professional development, and student loans for college education in an effort improve our science education programs. It is estimated that in that decade, two billion dollars were spent developing new approaches to teaching science and equipping science classrooms. It all began with the PSSC (Physical Science Study Committee) physics program for secondary science. Then, the impetus was carried into the elementary and middle school areas.

These curriculum projects were often labeled "teacher proof" to signify that even the least-prepared teacher could follow the directions and teach with them. This was foolish for several reasons. First, it was insulting to think that a teacher could not master new teaching techniques if given the proper training, and secondly, they ignored the fact that teachers had methods of teaching that could have and should have been modified, not merely dismissed. Just as children build on their preconceptions rather than discard them, teachers are capable of building on their prior instructional methods.

## Programs for Elementary Schools

### *The Days of the Alphabet Soup*

Due to the plethora of new programs and the propensity of the developers to assign acronyms, the many science curriculum projects formed what was called the "alphabet soup." SCIS, SAPA, ESS, HPP, PSSC, CHEMS, and ESCS were some of the programs that tried to make changes in the way science was taught. These were times of urgency and impatience to try new ideas. Scientists, psychologists, and educators were willing to collaborate on developing new curricula. Teachers were offered free summer courses at universities and colleges to upgrade their content background. They often came home with lots of scientific equipment and materials to be used in their classrooms, such as Geiger counters, cloud chambers, books, and other demonstration equipment.

One very popular program was and still is called Science Curriculum Instructional Strategy (SCIS) and is based upon the philosophy of Jean Piaget. It stressed science content in a reasonably efficient way in the biological and physical sciences. It moved children from the simplest conceptual areas to more complex understandings through activities using kit materials. Robert Karplus and Herbert Thier at the University of California at Berkeley's Lawrence Hall of Science developed the program, and it is still available commercially in modified form.

One of the major teaching models stressed in the program was the "learning cycle," which will be covered in more detail in Chapter 7. It consisted of three parts: concept exploration, concept introduction, and concept application. Children were allowed to explore in order to form questions and to revisit or reconstruct their prior knowledge. Then the teacher and students worked together to introduce a concept related to their exploration and, finally, that concept was applied to a new context to solidify the learning that took place.

According to some critics the problem was that the cycle of instruction relied too much on the teaching and little on the learning. We didn't know as much about learning in those days and focused instead on making teachers behave differently, which meant more hands-on activities and the use of many more materials. Many of the programs relied on rather expensive kits and replacing materials cut severely into school budgets, particularly after Congress tightened the purse strings compared to the early days of the NDEA, when every dollar spent was matched by the NSF (NRC 2007).

Another program (1960–1974) was the American Association for the Advancement of Science–sponsored Science: A Process Approach (SAPA) directed by Robert Gagne and Paul DeHart Hurd. This program was an attempt to focus on the processes of science taught across a broad spectrum of science content. The processes included observing, classifying, predicting, inferring, hypothesizing, and so on. Although it became quite popular in many school districts, it never achieved the market that made it effective in the larger population. The main criticism was that it separated content from process so that the content aspect of the program was compromised and lacked continuity. It also had the familiar kit and cost problem. However, its proponents insisted that students had many experiences using process skills in broad areas of the central concepts listed below:

- All matter is composed of units called fundamental particles.
- Matter exists in the form of units that can be classified into hierarchies of organizational levels.

- The behavior of matter in the universe can be described on a statistical basis; units of matter interact; and the basis of all ordinary interactions is electromagnetic, gravitational, or nuclear forces.
- All interacting units of matter tend toward the equilibrium states in which energy content is a minimum and the energy distribution is most random, and the sum of energy and matter in the universe remains constant.
- One of the forms of energy is the motion of units of matter. All matter exists in time and space, and, since interactions occur among these units, matter is subject, in some degree, to changes with time. (AAAS 1960)

Do you notice a similarity between this idea of science focusing on a few core concepts and the suggestions of fewer disciplinary core ideas in *A Framework for K–12 Science Education*?

At the same time, psychologists, educators and scientists at the Education Development Center (EDC) in Newton, Massachusetts, were developing the Elementary Science Study (ESS). David Hawkins had a great deal to do with the development of this program. He referred often to the term, "messing about," which is a phrase uttered by the Water Rat character from *Wind in the Willows*, the 1908 children's book by Kenneth Grahame. Hawkins wrote a famous article, paraphrasing the comments made by Rat, entitled "Messing about in Science," still considered a gem in science education literature (1965).

This program focused on pure inquiry and discovery with the premise that if children were presented with materials, the questions and inquiry would follow, given the leadership of qualified and competent teachers. The developers believed that science can be taught to children in a way that mirrors scientists' intellectual work in understanding the natural world. Several senior scientists were involved in the program from Harvard and the Massachusetts Institute of Technology (MIT) including Dr. Philip Morrison. There were three underlying assumptions that informed this program: (1) Scientific principles should be taught implicitly rather than explicitly; (2) children learn science best by doing science through hands-on learning that encourages a very open-ended approach with a minimum of teacher direction; and (3) lessons must take into account the research findings of developmental psychologists. Harvard psychologist Jerome Bruner had a strong influence on this program. ESS also varied from other programs in that it had trials both in the United States and in Africa. There were originally 56 different units developed, such as Pendulums, The Behavior of Mealworms, Batteries and Bulbs, and Mystery Powders. Thirty-eight are still in print and available from some distributors along with the materials and teacher guides.

These are just a few of the programs that were developed in the 1960s. We mention them here because although they were created before states developed standards for science, they have continued to influence elementary school science instruction in a variety of ways. Instructional materials developers have adapted their inquiry approach into their standards-based curriculum materials. Teachers today use many of the approaches pioneered in the early NSF-funded curricula, such as the learning cycle. Many of today's science education leaders began their own careers teaching these programs and some even developed or tested the original programs.

Those wishing to read more about the effect of the elementary science programs on education may refer to the research of James Shymansky and colleagues in their 1990 article in the *Journal of Research in Science Teaching* or do an internet search for "elementary science alphabet soup programs."

## Programs for Secondary Schools

While elementary science was undergoing these changes, programs were being developed for secondary science education as well. Here is a brief listing of some of the most important programs.

In 1950s, the Physical Science Study Committee, chaired by Professor Jerrold Zacharias of MIT, released a complete secondary physics program (PSSC Physics) that had a vast impact on physics classes in the United States (PSSC 1960). By the the early 1960s, about half of physics students in high schools were using this program. It continued to be in demand until about 1974. It actually was the prototype of activity-oriented programs and for the alphabet soup curricula that followed in its aftermath.

In 1958, with funding from the NSF, the American Institute of Biological Sciences, under the direction of Arnold Grobman and H. Bentley Glass, began work on the Biological Sciences Curriculum Study (BSCS), a biology program for high schools that was concept-based rather than fact-oriented and featured activities instead of lecture materials (AIBS 1958).

### Author Vignette

The BSCS product was ready for trials in 1959, and I was able to pilot some of the first versions. There were three "flavors," as teachers referred to them back then: blue, the molecular approach; orange, the human approach; and green, the ecological approach. I chose the molecular approach and began the first three weeks of the biology program teaching what seemed to be more like chemistry than biology. I remember students complaining that they thought that they were taking biology, but later, they admitted that the chemistry helped them understand the processes of biology.

As a result of a study I did years later with my British colleague John Leach, I became even more convinced that the particulate nature of matter approach was important to biological concept formation.

I had interviewed a young woman who had just completed her high school course in biology. I asked her if, when an apple decomposed, any of it would be available to supply material for the growth of another plant. She responded with an emphatic "No!" Then thinking a bit longer, she added, "Wait a minute, water recycles. (Pause) Ah, but not real stuff like apples."

During the same time period, I interviewed a young man who had not studied biology formally. He had taken chemistry instead and had no trouble seeing that the molecules and atoms of a decomposed organic substance could be used again by another plant. I believe that it was his belief in the particulate nature of matter that allowed him to answer in this way, and that the young woman had not considered this in her answer, even though she had completed a biology course. Although I did not have the opportunity to interview her biology teacher, I suspect the teacher did not believe that repeating the importance of the particulate nature of matter was necessary (Leach, Konicek, and Shapiro 1992).

One of the high school biology teachers in my school system became enamored with the ecological approach. The town was building a new high school, and this biology teacher used his version of the BSCS program to initiate an ecological study of the proposed school site and studied the impact of the building and its activities on the site for several years afterward.

—Dick Konicek-Moran

The BSCS program had a great impact on biology programs across the nation for years after its creation. Dick was also able to institute the new Earth Science Curriculum Project (ESCP), Investigating the Earth, around 1965 in the town's three middle schools. This program also focused on concepts and activities and was extremely timely due to the acceptance of such new ideas as plate tectonics. It was popular into the 1980s, when politicians began to question the nationally funded science curricula and their fitness for preparing students for the 21st century.

In the 1960s, after the success of the Russian Sputnik project, the NSF pressed for a new high school physics curriculum. At the suggestion of F. James Rutherford (then a graduate student under Fletcher Watson at Harvard), Gerald Holden of the Harvard Physics department agreed to head what was to become known as the Harvard Project Physics (later renamed Project Physics Course). Along with Watson, these three men developed a humanistic, historical-based curriculum text embellished with 8 mm movies, laboratory guides, and student guides. Over the next five decades it was modified and is known today as Exploring Physics. Its premise has always been that in order for students to enjoy the subjects in science curricula, they have to see science as the result of the work and lives of past scientists.

Around this time in the mid-1980s, most science instruction began to revert to its default position much as it had been before the post-"Sputnik" emphasis on change. Of course there were exceptions as teachers who had become devotees of the new science curricula continued to use the new programs.

## Analysis of the Early Programs

The programs discussed above had an impact on the direction of science education in elementary and secondary science during those decades, but each in its own way was lacking in its ability to convert teachers to a new way of looking at how children learn science or to convince textbook publishers that these ideas were saleable. Textbook publishers' decisions were certainly market driven. Some aspects of each program were incorporated into textbooks (the main source of instruction at the time), but without the entire philosophical background in place, the end result was disappointing. One main problem was that professional development had not progressed to an extent that teachers were trained sufficiently to carry out the programs as designed. Another problem was that the practices and content of science were seen by some as separate entities, a belief we no longer hold, as the *NGSS* do not separate content from the practices of science.

Yet, all of the programs had serious effects on the future of science education. And, in fact, they provided an impetus for change in science education that persists to this day. It is probable that these programs, analyzed as they were by researchers, prompted science educators to find a way to move to a more efficient way to reach the goals of science education. In subsequent years, new programs were developed, such as Full Option Science System (FOSS) developed at the Lawrence Hall of Science; Insights, developed by the Educational Development Corporation; Science and Technology for Children (STC) developed by the National Science Resources Center at the Smithsonian Institution; and Active Physics, written by distinguished educator and National Science Teachers Association past-president, Dr. Arthur Eisenkraft, to name a few.

These programs are still being used today and have made great inroads into the improvement of science teaching in schools. Many of these instructional programs were developed with funding from NSF's Instructional Materials Development Program. Each project was required to ally with a commercial distributor so that the expertise to get the products to the schools was employed. In addition, NSF provided funding for several implementation centers across the country that supported curriculum implementation once a program was adopted.

## Other Reform Movements and Projects Influencing Science Curriculum and Instruction

Project 2061 was formed in 1985 with far-reaching goals for science education. It is the reform arm of the AAAS.

> *Project 2061 began its work in 1985—the year Halley's Comet was last visible from earth. Children starting school now will see the return of the Comet in 2061—a reminder that today's education will shape the quality of their lives as they come of age in the 21st century amid profound scientific and technological change (www.project2061.org).*

Project 2061 is responsible for such publications as the *Atlas of Science Literacy,* volumes 1 (AAAS 2001) and 2 (AAAS 2007), *Benchmarks for Science Literacy* (AAAS 1993, 2009), and *Science for All Americans* (AAAS 1989). Project 2016 also conducted an in-depth analysis of middle and high school science textbooks and showed that while "heavy for their size, they were light on learning" (Roseman, Kesidou, Stern, and Caldwell 1999). Their analysis raised concerns about the quality of textbooks in terms of supporting learning and led to several major revisions in the textbook publishing industry. A description of Project 2061's high school science textbook analysis procedure and findings can be viewed at *www.project2061.org/publications/textbook/hsbio/report/default.htm.*

## *A Private Universe* and Alternative Conceptions

Research was finding evidence that there was vast room for improvement in science education. Studies showed that students came to the classroom with more information about science phenomena than previously suspected. Researchers were realizing that children are more capable of engaging in sophisticated thinking, and the spirit of inquiry that Einstein wished for was not finding its way into the classroom.

Some educators would say that one of the most effective catalysts in changing the ways science was taught in the 1990s was the introduction of the many online programs put forth by the Science Media Group of the Harvard-Smithsonian Center for Astrophysics in Cambridge, Massachusetts, led primarily by Drs. Matthew Schneps, Irwin Shapiro, and Philip Sadler. Arguably, their most influential production was the video series *A Private Universe,* which opened the eyes of the nation to the fact that children and adults often leave educational institutions with the same preconceived ideas about the natural world as they had when they entered (1987).

## Author Vignette

Both Dick and I were strongly influenced by the *A Private Universe* Project. Dick was directly involved with the project and appeared in several of the videos. I was a middle school teacher in the early 1990s, participating in an Massachusetts Corporation for Educational Telecommunications Stars Schools program that brought live, interactive television into schools. *A Private Universe* was broadcast as a live interactive satellite program in which teachers gathered after school for professional development and interaction with the program. It was the first time I had considered the impact of students' prior knowledge and everyday experiences on teaching and learning and how they can pose critical barriers to understanding. I will always remember Dick explaining how a constructivist approach to teaching can be used to build a bridge from where the student is to where the student needs to be in their scientific thinking. It definitely changed my approach to teaching and has had a lingering effect on my work today in developing formative assessment probes that teachers can use to uncover their own students' "private universe." I continue to use the Annenberg videos in teacher professional development, especially to help teachers recognize that even our brightest students bring alternative ideas to their learning and that effective teaching involves uncovering these ideas and helping students work through them so they willingly give them up in favor of the correct scientific ideas.

—Page Keeley

This video and the many online seminars produced by the Science Media Group may well have initiated the focus of the next decade of science education research: children's pre-instructional- and instructional-resistant ideas, commonly labeled *misconceptions*. The resilient nature of these "naive" ideas taught us two things that were very influential in the paradigms that dominate science education even today: (1) Children come to school with preconceived ideas about the way the world works, and (2) these ideas are very resistant to change.

At approximately the same time, Bonnie Shapiro, in her book *What Children Bring to Light*, documented how some teachers could be lulled into believing that their students understood science concepts while still holding on to their prior conceptions (1994).

Teachers were introduced to ways of finding out what their students were thinking, to use techniques to move the students toward scientific knowledge more in tune with the scientific community, and to provide starters for science talks in classrooms. It was the beginning of an era where there was an alternative to summative evaluations and questions at the end of the chapter that used to be the standard option open to teachers in their preparation of lessons for children. Obviously, it was time for a paradigm shift in science education.

## Early Standards of Development and Cognitive Research

In 1993, AAAS published its companion opus *Benchmarks for Science Literacy* (updated in an online version in 2009). This was the first set of "learning goals" in science, mathematics, social science, and technology established through consensus by a broad constituency of scientists and science education stakeholders. The *Benchmarks* informed the development of many states' first curriculum frameworks and standards. Shortly after the *Benchmarks,* in 1996, the National Academy of Science's National Research Council produced the *National Science Education Standards* (*NSES*). Their focus was on inquiry-based science based strongly on the theory of constructivism instead of on direct instruction of facts and methods. The *NSES* emphasized *inquiry* as the key to science learning. When engaging in inquiry, students were expected to describe objects and events, ask researchable questions, construct explanations, test those explanations against current scientific knowledge, and be able to communicate their ideas coherently. They were expected to identify their claims, use critical and logical thinking, and consider alternative explanations. It was expected that students would develop their understanding of science by combining scientific knowledge with investigative and reasoning and skills.

Unfortunately, as is often the case, this did not happen as well as expected in the average classroom. Children did not develop the understanding of "standards-based" science concepts as evidenced in the Trends in International Math and Science Study (TIMSS) test given to fourth and eighth graders all over the world every four years since 1995, when U.S. students achieved well below children in other countries.

More current research into how humans learn and specifically how children learn led to the next step in improving science education. The book *How People Learn,* published by the National Research Council (NRC), compiled the latest research on learning and became the go-to book for those interested in a cognitive approach to science education (Bransford, Brown, and Cocking 2000). *How People Learn* emphasized the importance of taking the time to find out the preconceptions students bring to their learning and examined four environments that support science learning: learner centered, knowledge centered, assessment centered, and community centered. It characterized the nature of expertise from novice to expert and described challenges for educators to be aware of in teaching novice learners. It described three important implications for teaching: (1) making thinking visible, both student thinking and expert thinking; (2) being aware of the knowledge level of students; and (3) using contrasting cases to highlight a point or set of points. There was such widespread interest in learning about learning that Harold Pratt, during his NSTA presidency, commissioned an essay collection edited by Rodger Bybee called *Learning Science and the Science of Learning* (Bybee 2002). In this book, noted science educators discuss research findings on how students and teachers learn and how to translate those findings into practical classroom applications.

## *A Framework for K–12 Science Education* and the *Next Generation Science Standards*

The NRC was given the task of producing a new K–12 science education framework based on recent research and their assessment of the weaknesses of existing standards. The *Framework* would inform the development of a new, updated set of science standards. These *Next Generation Science Standards* (*NGSS*) would focus on core disciplinary ideas, crosscutting concepts, and scientific and engineering practices all students should learn by the end of high school. It would also include performance expectations that combine content and practice. It would hopefully do away with the oft-criticized curriculum: the one often referred to as "a mile wide and an inch deep" or the "M&M" curriculum (Mention and Move On). The *NGSS* would focus on fewer "core concepts" over the entire K–12 school experience and would feature an emphasis on scientific and engineering practices that all students would experience and connections to unifying ideas called crosscutting concepts. These were developed with input from lead states and available to the public for review in 2012 before being published in 2013. Basic features of the *NGSS* include

- performance expectations for all students;
- a progression of learning from K to 12;
- clear and coherent learning goals, but not a curriculum;
- promotion of excellence and equity;
- integration of engineering with science; and
- preparation for college and career.

It is beyond the scope of this chapter to go into details about the *Framework* and the *NGSS*. Much of this information can be accessed through the NSTA *NGSS* Hub at *http://ngss.nsta.org*. What we find promising about the *NGSS* is that we now have a set of standards that will move educators beyond an overwhelming number of disconnected pieces of knowledge and skills that often get treated like a checklist to standards that will encourage teaching for deeper and a more coherent conceptual understanding.

In the next chapter, we will consider how the scientific and engineering practices included in *A Framework for K–12 Science Education* and the *Next Generation Science Standards* have the potential to transform the teaching and learning of science for conceptual understanding in our schools.

## Author Vignette

With all of the published background information on *A Framework for K–12 Science Education* and the *Next Generation Science Standards*, I don't think many science educators are aware of the initial role NSTA played in getting this effort started. When I was NSTA President-Elect in 2007, Dr. Gerry Wheeler, Executive Director of NSTA at that time, proposed the idea of a new NSTA-led initiative, the Science Anchors project. Dr. Wheeler explained that science standards were too broad and too many, resulting in superficial coverage that fails to link science concepts and develop them over successive grade levels. He proposed an initiative that would identify and focus on a core set of important ideas in science. The Science Anchors project would convene a group that would look at existing national standards, international benchmarks, and state standards to identify a core set of ideas. An NSTA task force made up of distinguished leaders in science education, and co-chaired by Dr. Cary Sneider and me, convened in Chicago to brainstorm ideas about how to undertake this ambitious initiative. A three-dimensional "abacus" model of core ideas, scientific and engineering practices, and crosscutting concepts was described. The group strongly encouraged the use of learning progressions and current research to inform grade-level placement of core ideas. It was also felt that this was an undertaking that NSTA could not do alone and warranted the cultivation of a collaborative group to go forward with this idea. The task force report was submitted to the NSTA Board of Directors in 2009 and approved.

During my NSTA Presidency in 2008–09, we met with several groups to consider how to best go about the board-approved Science Anchors project. We met at NSTA headquarters with representatives from AAAS/Project 2061, the National Research Council, and Achieve. Focus groups were held at all of the NSTA conferences to seek input from stakeholders. Ultimately the NRC took on the task of convening a select panel to develop *A Framework for K–12 Science Education*; Achieve took on the task of working with lead states to write the standards;

> and NSTA encouraged its members to review drafts, led focus groups, supported dissemination, featured *NGSS* sessions at its conferences, and produced webinars and support materials.
>
> —Page Keeley

## Questions for Reflection and Discussion

1. Re-read John Dewey's quotes on pages 79–80 of this chapter. What is your opinion of what he has to say, particularly in the third quote? If you could have a conversation with John Dewey today about science education, what would you talk about?
2. Search on the internet for further information on the "alphabet soup" programs described in this chapter. What aspects of these programs do you think still exist in science curriculum programs today? What aspects could be modified for use with today's new standards?
3. Visit the Project 2061 website (*www.project2061.org*) and examine the science education reform tools and resources they have produced. How are these tools and resources still useful today?
4. View videos of *A Private Universe* (or *Minds of Their Own*) available on the Annenberg Learner website (*www.learner.org*). How are these videos still useful today in helping teachers learn more about effective teaching?
5. What will be the biggest challenge for you as science education transitions to new standards, such as the *NGSS*? How will you address these challenges?
6. Do you think improving science education is a never-ending process? Why or why not? How can examining the past help inform our thinking about the present and the future.
7. Recently, an educational expert stated that the new standards will suffer the same fate as the old ones: leading teachers to teach to the test and teach test-taking skills. How do you feel about this and what would you do to try to avoid this pitfall?
8. Choose one "golden line" from this chapter (a sentence that really speaks to or resonates with you). Write this on a sentence strip and share it with others. Explain why you chose it.
9. What was the biggest "takeaway" from this chapter for you? What will you do or think about differently as a result?

## Extending Your Learning With NSTA Resources

1. NSTA has an *NGSS* Hub that provides extensive resources for learning about and implementing the *NGSS*. Visit *http://ngss.nsta.org* to access these resources, including books, journal articles, *NGSS* charts, web seminars, and more.
2. Renowned educator Rodger Bybee provides a unique perspective on reflecting on the past and thinking about the future of science education in Bybee, R. 2010. *The teaching of science: 21st-century perspectives.* Arlington, VA: NSTA Press.
3. Since 2012, the NSTA journals *Science and Children, Science Scope,* and *The Science Teacher* have included monthly guest editorials and articles on the *Framework* and *NGSS*. Check the online archives of all journals published since 2012 to access and read these articles. These articles can be accessed for free using your NSTA membership number at *www.nsta.org/publications/#journals*. Many articles can also be accessed through the NSTA Learning Center at *http://learningcenter.nsta.org/products/journals.aspx*.

# Chapter 6

## How Is Conceptual Understanding Developed Through the Three Dimensions and Learning Strands?

In 2012, the NRC published *A Framework for K–12 Science Education: Practices, Crosscutting Concepts, and Core Ideas.* The purposes of the document were twofold: first, to elaborate a clear and coherent set of expectations for K–12 science students and their teachers; and second, to provide a guide for the writers of the *Next Generation of Science Standards* (*NGSS*). It offers guidelines for curriculum developers and teachers in science and engineering programs in elementary and secondary schools. Paving the way for this document was a research report, *Taking Science to School* (NRC 2007) that described how children learn science. Much of the *Framework* was based on the suggestions from this report, as well as from the practitioner's version, *Ready, Set, SCIENCE!* (Michaels, Shouse, and Schweingruber 2008). Additionally, the *Framework* also builds on the three prior national works on standards: *Benchmarks for Science Literacy* (AAAS 2009), the *National Science Education Standards* (NRC 1996), and *College Board Standards for College Success in Science* (College Board 2009). Many things have changed and lessons have been learned over the almost two decades since *Benchmarks for Science Literacy* and the *National Science Education Standards* were released, yet some things are very similar.

> *Several guiding principles, drawn from what is known about the nature of learning science, underlie both the structure and the content of the framework. These principles include young children's capacity to learn science, a focus on core ideas, the development of true understanding over time, the consideration both of knowledge and practice, the linkage of science education to students' interests and experiences, and the promotion of equity. (NRC 2012, p. 24)*

Based on these guiding principles, a framework was developed made up of three dimensions that outline the knowledge and practices of K–12 science and engineering. These three dimensions, which are woven together to develop learning goals, are

- the core ideas in the science disciplines and the relationships among science, engineering, and technology;

- the crosscutting concepts; and
- the scientific and engineering practices.

## Core Ideas

State standards and curriculum have been criticized because too many topics are "covered," at the expense of developing deep and coherent understanding. Covering, rather than *uncovering* of knowledge, seems to be the focus in some classrooms and school districts, except in those pockets of innovation where teachers are given more autonomy to do what is best for student learning. In a study that included over 8,000 high school graduates in some 55 different colleges, students who reported they had studied at least one major topic in depth did significantly better in college courses than those that covered the entire curriculum and studied no topic in depth (Schwartz, Sadler, Sonnert, and Tai 2009).

### Author Vignette

In the early 1990s, I had the opportunity to work with the Harvard-Smithsonian Center for Astrophysics on the Private Universe Project, which in addition to producing the video *Minds of Their Own* (Harvard-Smithsonian Center for Astrophysics 1997) also produced a teacher workshop series through interactive TV (Annenberg 1995). In workshop #3 "Hands-On/Minds-On Learning," a veteran high school teacher named Mr. Carter was featured. Mr. Carter is quite surprised to see that many of his physics students misinterpret his well-planned lab activities and fail to grasp some of the fundamental ideas, such as how to light up a bulb with a battery and a wire. We paired him with physics education researcher and teacher Jim Minstrell, who helped Mr. Carter redefine his role as a teacher, taking time to elicit students' thinking and teach in greater depth. Much to his surprise, his students liked the "new Mr. Carter" as he helped his students understand the workings of batteries and bulbs before going into the details of circuitry. When Mr. Carter asked his students to respond to the fact that this method of teaching and learning took more time, a student answered: "Say you teach us something and some of us don't get it and you go on to something else. Isn't that bad? Shouldn't we learn that section first and take more time and learn that better? Even if we don't finish that chapter or the book, or whatever?"

—Dick Konicek-Moran

Focusing on core ideas and teaching for depth is important for developing conceptual understanding. With so much information available at our fingertips today, the emphasis has moved from teaching a plethora of facts to focusing on the core knowledge students need so they can acquire the additional information they need to understand the natural world and make decisions based on knowledge of science and technology.

The disciplinary core ideas were carefully selected by the National Research Council (NRC) committee to focus K–12 science curriculum, instruction, and assessments on the most important aspects of science. They are grouped into four domains: physical sciences; life sciences; Earth and space sciences; and engineering, technology, and applications of science. To be considered a core idea, the NRC committee identified four criteria. Core ideas had to meet at least two of the following criteria and preferably three or all four (NRC 2013):

- Have *broad importance* across multiple sciences or engineering disciplines or be a *key organizing concept* of a single discipline.
- Provide a *key tool* for understanding or investigating more complex ideas and solving problems.
- Relate to the *interests and life experiences of students* or be connected to *societal or personal concerns* that require scientific or technological knowledge.
- Be *teachable* and *learnable* over multiple grades at increasing levels of depth and sophistication.

Some educators may find their "favorite" science concepts are not included in the smaller set of disciplinary core ideas, or that some ideas in the standards adopted by their states are not reflected in the disciplinary core ideas. *The Opportunity Equation* report by the Carnegie Corporation and the Institute for Advanced Study (IAS) concluded that the *National Science Education Standards* (*NSES*), used by many states to develop their standards as well as inform the development of instructional materials, may have even contributed to the "mile-wide, inch-deep" curriculum (Carnegie-IAS 2009). The committee identified that the set of core ideas could not include everything that is typically taught in the K–12 science curriculum. If the goal is to teach to standards that are fewer, clearer, and higher, then thoughtful, research-informed decisions needed to be made about what a strong base of core knowledge, conceptually understood in sufficient depth, would include (and exclude) so students would complete a K–12 education well-grounded in science and able to make connections across concepts. As a result, K–12 educators now have the guidance to focus on conceptual understanding of scientific concepts and ideas, rather than covering many disconnected concepts and ideas that are often memorized for a test and then quickly forgotten.

## Crosscutting Concepts

The idea of concepts that cut across all the disciplines of science and that all science-literate people should know and be able to use was first addressed in *Science for All Americans* (AAAS 1988). Called the *common themes* of science, they were described as follows: "Some important themes pervade science, mathematics, and technology and appear over and over again, whether we are looking at an ancient civilization, the human body, or a comet. They are ideas that transcend disciplinary boundaries and prove fruitful in explanation, in theory, in observation, and in design" (AAAS 1988, p. 165).

*Benchmarks for Science Literacy* (AAAS 2009) includes explicit learning goals and instructional considerations for four crosscutting concepts, or *common themes* as they are referred to in the *Benchmarks*: systems, models, constancy and change, and scale. For example, students in K–2 should begin to use the concept of *systems* to identify how one thing affects another in order to lay the groundwork for recognizing interactions, and practice identifying parts of things and how one part connects to and affects another. In grades 3–5, students use the concept of *models* to suggest how models are like and unlike the real thing and identify limitations of models. In middle school students use the concept of *constancy and change* to look for patterns, including rates of change and cyclic patterns and variation. In high school, students use the concept of *scale* to examine how models that work well on one scale may not work well or at all if the scale is greatly increased or decreased.

Building upon the common themes from the *Benchmarks* and the unifying concepts and processes in the *National Science Education Standards, A Framework for K–12 Science Education* and the *Next Generation Science Standards* include similar "big ideas" referred to as the crosscutting concepts. Like their predecessors, these concepts bridge disciplinary boundaries and have explanatory value in helping students understand scientific ideas. Regardless of the labels or organizational schemes used in these documents, all of them stress that it is important for students to come to recognize the concepts common to so many areas of science and engineering. The seven crosscutting concepts are

- Patterns;
- Cause and effect;
- Scale, proportion, and quantity;
- Systems and system models;
- Energy and matter: Flows, cycles, and conservation;
- Structure and function; and
- Stability and change.

Why are crosscutting concepts important for developing conceptual understanding? They provide an organizational schema for connecting knowledge from the physical, Earth, space, and life sciences into a coherent and scientifically based view of the natural world.

## Scientific and Engineering Practices

"Where is *inquiry* in all this?" you might ask. After all, we have been advocating for inquiry for the past several decades. Rest assured, inquiry has not been forgotten, but it has become evident that inquiry consists of many things, and the scientific and engineering practices found in *NGSS* include the parsed-out elements of inquiry that mirror the way science is practiced in the real world, not just in school science. If we view a classroom as analogous to the scientific community, the practices point the way to the behaviors, processes, and habits of mind we want to stress for science to become a natural way of thinking and understanding the natural world. The practices actually provide a very clear description of what teaching for inquiry should look like. The practices are not a set of process skills taught in isolation of the content to be learned, but rather they describe what scientists actually do to investigate the natural world to build understanding. Students should not only be able to use the practices but also conceptually understand a practice to fully appreciate the nature of scientific knowledge.

For quite some time, science educators believed that "hands-on" activities were the answer to children's understanding through their participation in science-related activities. Many teachers believed that students merely engaging in activities and manipulating objects would organize the information to be gained and the knowledge to be understood into concept comprehension. Educators began to notice that the pendulum had swung too far to the "hands-on" component of inquiry as they realized that the knowledge was not inherent in the materials themselves, but in the thought and metacognition about what students had done in the activity. We now know that "hands-on" is a dangerous phrase when speaking about learning science. The missing ingredient is the "minds-on" part of the instructional experience. Clarity about the knowledge intended in any activity comes from each student's re-creation of concepts—and discussing, thinking, arguing, listening, and evaluating one's own preconceptions after the activities, under the leadership of a thoughtful teacher can bring this about. After all, a food fight is a hands-on activity, but about all you would learn was something about the aerodynamics of flying mashed potatoes! Our view of what students need to build their knowledge and theories about the natural world extends far beyond a "hands-on activity." While it is important for students to use and interact with materials in science class, the learning comes from the sense-making of students' "hands-on" experiences.

Research (e.g., Brice-Heath 1983; Lemke, 1990; Michaels and Sohmer 2001) now shows that the more the students talk among themselves, examine their thinking, and engage in dialogue with their fellow students and teachers, the more they understand the meaning of the activities in which they engage. "Truth springs from arguments between friends," said 18th-century Scottish philosopher David Hume. Teachers used to caution students to work silently and not talk to their neighbors. This is no longer so, since talking and listening to others is touted as one way to better understanding. Some principals we have talked with call it "productive noise."

Below is the list of the scientific and engineering practices, as defined in the *Framework* and *NGSS*. We will describe each of them in some detail, focus primarily on the scientific practices, and make the connections to teaching for conceptual understanding:

1. Asking questions and defining problems
2. Developing and using models
3. Planning and carrying out investigations
4. Analyzing and interpreting data
5. Using mathematics and computational thinking
6. Constructing explanations and designing solutions
7. Engaging in argument from evidence
8. Obtaining, evaluating, and communicating information

## Asking Questions and Defining Problems

*A Framework for K–12 Science Education* summarizes this science practice this way:

> *Science begins with a question about a phenomenon, such as "Why is the sky blue?" or "What causes cancer?" and seeks to develop theories that can provide explanatory answers to such questions. A basic practice of the scientist is formulating empirically answerable questions about phenomena, establishing what is already known, and determining what questions have yet to be satisfactorily answered. (NRC 2012, p. 50)*

Recall our earlier reference to Nobel Laureate Isadore Rabi and his mother's query each day after school, "Izzy, did you ask any good questions today?" Asking questions is the driving force of science. Defining problems is the driving force of engineering. Both stem from a need to find a researchable question that needs to be answered or a problem solved.

Children enter school with a myriad of questions, as described by Karen Gallas in her book *Talking Their Way Into Science: Hearing Children's Questions and Theories, and Responding With Curricula* (1995). Gallas, a primary teacher, finds that children have many questions about their world—a possible world that often ends as a story that explains real life. Asking and answering questions in a classroom can be all about "power." The teacher who prepares a lesson for her students is often unaware of the "power" she holds about which questions from the students are answered and which are ignored.

An article by Beck and Leishman (1996) describes how a teacher unknowingly ignores a very sincere question from a student, Daniel, who asks, "Can the flower drink green water?" The principal (Beck) is observing a teacher (Leishman) as she teaches a lesson on how plants obtain nutrients and water. The principal, in the interview after the observation, calls to the teacher's attention that she ignored Daniel's question. The teacher is aghast that she not only ignored the question but that she did not even remember the question being asked. She says:

> *After this experience, I was left with some unsettling and important questions. It is frightening how much power I have in the classroom. I decide which questions are worthy of being pursued and which ones get a slight nod of the head, indulgent smile, no response, or even a reprimand. (p. 57)*

When asked if she remembered the question, she wrote:

> *In hindsight, Daniel's question fit none of the boxes I had already determined went with this lesson. I was not prepared to discuss anything that detracted from the lesson on how the water gets from the root to the flower. I knew the answer to his question: of course it will drink green water. The food coloring had not changed the water. Therefore, the question was unworthy, in fact, I thought, a silly question. (p. 58)*

Later, she found out that not only did Daniel believe that green water was undrinkable but that many of his fellow students were of the same impression. She designed a new lesson that addressed the conceptions that troubled the children: that somehow, food coloring in water could be sensed by the plant and be refused entrance. She then wrote about her learning from the incident. Especially telling is Leishman's realization below about children's questions being a *window into their understanding:*

> *I am the biggest learner in this group. I learned that children's questions are my window into their understanding, thinking, and perceptions and are much more telling than any pretest I could give. I learned that I need to be extremely thoughtful before dismissing questions as unworthy. I need to make sure that I do not let my own ignorance on a subject dictate what questions I choose not to address. (Beck and Leishman 1996, p. 59)*

According to Karen Gallas, it is not too difficult to encourage children to ask questions, if the tone of the classroom is properly set. The difficulty may be in getting them to ask questions that are empirically answerable about things that are important to them. The mystery stories in the *Everyday Science Mysteries* series or the formative probes in the *Uncovering Student Ideas in Science* series are helpful in setting the stage for this practice. In any case, it is important not only to help students see that questions they raise can be answered in many ways but to make sure that their questions are related to the children's lives.

*NGSS* Appendix F (Science and Engineering Practices in the *NGSS*) includes a matrix for examining the progression of what students are expected to do for each practice at different grade spans. The matrix for Practice 1: Asking Questions and Defining Problems, is in Table 6.1 (p. 100). As you examine how students use this practice, notice how it overlaps with other practices. (*Note:* While this practice includes the engineering practice of Defining Problems, for the purpose of this book, we are focusing primarily on the scientific practice of Asking Questions.)

## Table 6.1. Practice 1: Asking Questions and Defining Problems

**Practice 1: Asking Questions and Defining Problems**

| Grades K–2 | Grades 3–5 | Grades 6–8 | Grades 9–12 |
|---|---|---|---|
| Asking questions and defining problems in K–2 builds on prior experiences and progresses to simple descriptive questions that can be tested.<br>• Ask questions based on observations to find more information about the natural and/or designed world(s).<br>• Ask and/or identify questions that can be answered by an investigation.<br>• Define a simple problem that can be solved through the development of a new or improved object or tool. | Asking questions and defining problems in 3–5 builds on K–2 experiences and progresses to specifying qualitative relationships.<br>• Ask questions about what would happen if a variable is changed.<br>• Identify scientific (testable) and non-scientific (non-testable) questions.<br>• Ask questions that can be investigated and predict reasonable outcomes based on patterns such as cause and effect relationships.<br>• Use prior knowledge to describe problems that can be solved.<br>• Define a simple design problem that can be solved through the development of an object, tool, process, or system and includes several criteria for success and constraints on materials, time, or cost. | Asking questions and defining problems in 6–8 builds on K–5 experiences and progresses to specifying relationships between variables and clarifying arguments and models.<br>• Ask questions<br>o that arise from careful observation of phenomena, models, or unexpected results, to clarify and/or seek additional information.<br>o to identify and/or clarify evidence and/or the premise(s) of an argument.<br>o to determine relationships between independent and dependent variables and relationships in models.<br>o to clarify and/or refine a model, an explanation, or an engineering problem.<br>o that require sufficient and appropriate empirical evidence to answer.<br>o that can be investigated within the scope of the classroom, outdoor environment, and museums and other public facilities with available resources and, when appropriate, frame a hypothesis based on observations and scientific principles.<br>o that challenge the premise(s) of an argument or the interpretation of a data set.<br>• Define a design problem that can be solved through the development of an object, tool, process, or system and includes multiple criteria and constraints, including scientific knowledge that may limit possible solutions. | Asking questions and defining problems in 9–12 builds on K–8 experiences and progresses to formulating, refining, and evaluating empirically testable questions and design problems using models and simulations.<br>• Ask questions<br>o that arise from careful observation of phenomena, or unexpected results, to clarify and/or seek additional information.<br>o that arise from examining models or a theory, to clarify and/or seek additional information and relationships.<br>o to determine relationships, including quantitative relationships, between independent and dependent variables.<br>o to clarify and refine a model, an explanation, or an engineering problem.<br>• Evaluate a question to determine if it is testable and relevant.<br>• Ask questions that can be investigated within the scope of the school laboratory, research facilities, or field (e.g., outdoor environment) with available resources and, when appropriate, frame a hypothesis based on a model or theory.<br>• Ask and/or evaluate questions that challenge the premise(s) of an argument, the interpretation of a data set, or the suitability of a design.<br>• Define a design problem that involves the development of a process or system with interacting components and criteria and constraints that may include social, technical, and/or environmental considerations. |

*Source:* Appendix F: Science and Engineering Practices in the *NGSS*. NGSS Lead States 2013, p. 51.

### *Developing and Using Models*

*A Framework for K–12 Science Education* summarizes this science practice this way:

> *Science often involves the construction and use of a wide variety of models and simulations to help develop explanations about natural phenomena. Models make it possible to go beyond observables and imagine a world not yet seen. Models enable predictions of the form "if … then … therefore" to be made in order to test hypothetical explanations. (NRC 2012, p. 50)*

We have already touched on the nature of models in science education in Chapter 1. Be they physical, mental, mathematical, or theoretical, they all share the commonality of being analogous to a real situation or phenomenon. They make our thinking visible, provide explanatory power, and can be used to test predictions. Model-based instruction encourages students to make their thinking visible by drawing, writing, using a computer, or any other appropriate method so the model can be discussed, critiqued, and modified if necessary so the classroom community can reach a consensus on its viability.

## Author Vignette

In Amherst, Massachusetts, a few hundred yards from my house, there was a corner where four roads came together. None of them met at the same place, so anyone travelling on any of them could not see oncoming traffic. It was a traffic nightmare and caused many accidents. Engineers saw the problem as one that required that traffic could flow safely in any direction without delay. Naturally, the engineers could not build actual solutions and try them out, so they created a computer model with variables that could be manipulated, and it turned out that two rotaries allowed the best model for sustained traffic flow. Many people were not convinced that two rotaries were necessary to achieve the goals of continuous traffic flow and safety but upon seeing the results on the computer model, they were converted. The Commonwealth of Massachusetts was so convinced that the model was correct that it invested over 8 million dollars to construct the real thing. There is little doubt that without the computer model, the state would have considered the project a fool's errand. Although the affected community drolly refers to the two rotaries as "fruit loops," there is no doubt that the change is effective. Traffic moves efficiently, is slower and less dangerous, and we no longer hear the screech of tires in the middle of the night.

—Dick Konicek-Moran

### Creating and Changing Models

Metacognition (thinking about one's thinking) is connected closely to model-based instruction. If we are to use the ideas proposed by the authors of *Taking Science to School*, we think of models as a way of making our thoughts visible or at least public (NRC 2007). Phil Scott of the Children's Learning in Science Project (CLIS) at Leeds University in the United Kingdom was partial to using posters made by the students to illustrate their mental models of concepts (1994, personal conversation).

Children use models all of the time in their play. A stick becomes a gun or a sword. In classrooms, blocks become representations of candy bars used to solve mathematical problems. Young children have fascinating imaginations and can accommodate their thoughts into any object and make it into whatever they wish. Models can be the center of instruction if children are encouraged to build a mental or physical model that depicts their interpretation of a real object (circulatory system) or a phenomenon (water evaporating from a puddle). The children thus make their thoughts visible and can modify the model after testing it or validating its authenticity.

John Clement and his research group at the University of Massachusetts Amherst have done a great deal of research on the topic of *analogy,* a type of conceptual model. One of the main points that his group espouses is that modification of models is an evolutionary process, moving from the original naive model to the target model. This means that one cannot always expect to move to the target model in one step. Just as in evolution in the biological world (a metaphor), we expect to make small changes in the model, each of which is reviewed, tested, and revised (by natural selection in nature and by the classroom community in schools) until the students are ready to make the next step. Interventions by the teacher are meant to cause dissension in the students' thinking to provide bases for criticism and revision. Clement also stresses the need for teacher-student coconstruction of new models (2008).

For example, when middle or high school students are asked to describe the forces acting on a book resting on a table, they usually identify the gravitational force pulling on the book, but fail to recognize there is also a force, called the normal force, pushing upward on the book. Students have a hard time conceptualizing that a table pushes on a book at rest. To make the analogy, teachers can have students push on a spring. They feel the downward force of their hand pushing down on the spring, but they also feel the spring push up on their hand as it deforms. The teacher might suspend a meterstick between two tables and place a book on the center of the meterstick. The students can see the meterstick bend to accommodate the book at rest. The students relate this analogy to the spring. The teacher then places a second meterstick on top of the first meterstick to add rigidity, and again places a book in the center so it is resting on the metersticks. This time the students can barely detect a bend in the metersticks, but from the previous examples they know that the metersticks also push up on the book. From this bridging analogy, they can now make the leap to the book resting on the table and identify the pair of forces, including the force exerted by the table, on the book.

One might ask us what the difference is between an analogy and a model. It was interesting to us that when we researched this question, we were referred to philosophical journals or articles. Philosophers have argued questions like this for centuries. But we believe that an analogy becomes a model when it explains something to one's satisfaction. In science, we very often use models to research questions when the reality of the real world is unattainable. This is usually when the objects approach infinity on either side of the size scale, atoms and the cosmos, for example.

To look at a different example, Ohm's Law of electrical phenomena was developed from an analogy. This analogy looked at the flow of electricity as a fluid moving through a narrow pipe, with force moving the fluid along and resistance being the friction from the walls of the pipe. There were enough similarities in this analogy to continue its use for these many years. What became Ohm's Law was a model that works in ideal situations.

It is difficult, if not impossible, to find ideal situations that correspond to an analogy. Each analogy and model has its limitations and must be modified when there is need. The public often sees analogies and models as ideal and unquestioned. This can cause a problem, especially when it leads to definitions that can obstruct modification in one's mind. The vignette below shows how a definition of a liquid can cause a problem in conceptual change (*Note:* The full vignette is described in Chapter 26, "Is It a Solid? Claim Cards and Argumentation" of *What Are They Thinking?* [Keeley 2014]).

## Author Vignette

I had an opportunity to work with a fifth-grade teacher who was planning a series of lessons to help her students understand that all matter is composed of tiny particles we cannot see and that using a particle model can help us understand how matter behaves. In preparation for her lessons on gases, she decided to use a formative assessment probe to find out what her students already knew about the other two states of matter—solids and liquids.

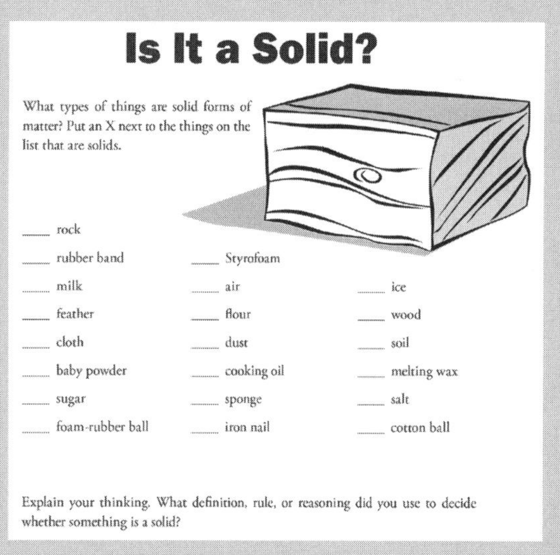

**Is It a Solid?**

What types of things are solid forms of matter? Put an X next to the things on the list that are solids.

___ rock
___ rubber band
___ milk
___ feather
___ cloth
___ baby powder
___ sugar
___ foam-rubber ball
___ Styrofoam
___ air
___ flour
___ dust
___ cooking oil
___ sponge
___ iron nail
___ ice
___ wood
___ soil
___ melting wax
___ salt
___ cotton ball

Explain your thinking. What definition, rule, or reasoning did you use to decide whether something is a solid?

She used the formative assessment probe "Is It a Solid?" in which students selected items from a list they thought were solids and explained their reasoning (Keeley, Eberle, and Dorsey 2008). The teacher transcribed a part of the students' discussion:

**Kara:** "My claim is that wood is a solid. My evidence is that a block of wood keeps its shape, and if you measure the amount, it stays the same."

**Max:** "But you can change its shape by cutting it."

**Kara:** "Yes, but whatever shape you cut it in, it holds that shape. It doesn't spread out or anything."

**Freddie:** "Yeah, solids are hard, and wood is hard, so I agree with the claim."

**Teacher:** "Does anyone want to add to or disagree with Kara's claim? Hearing no other ideas, let's for now put wood under solid, on our claims chart. Who would like to go next?"

**Ivy:** "I have flour. My claim is flour is not a solid. It is a liquid. First I thought it might be a solid because it isn't wet like a liquid."

**Teacher:** "Ivy, you said it isn't wet like a liquid, but you claim it is a liquid. Can you tell us why you think it is a liquid?"

**Ivy:** "Well, it's like what we've learned about liquids. You can pour them out and when you put them in something, they filled the shape of the thing you put them in like water does when you put it in a cup. You can pour it and the water fills out the cup. So I think flour does the same thing."

**Teacher:** "Sharla, you want to add something?"

**Sharla:** "I agree with Ivy. When me and my sisters make cookies, we pour the flour, and it's not hard like wood is."

**Teacher:** "Who else has an idea to support or disagree with Ivy's claim?"

**Hector:** "I'm not sure. It's kinda like water but not really, so I think it might be a solid."

**Pete:** "But it doesn't keep the same shape like wood."

**Teacher:** "We seem to have two different ideas about flour—some of you think it is a solid, some think it is a liquid, and others are not sure. Let's take a vote on where to put it on our claims chart. Remember, we will go back to our claims chart after our discussion, and you will have a chance to change your claims after you hear more arguments about whether the things on your cards are solids or not solids."

Class vote indicates most think flour is a liquid, so the teacher lists flour under "not a solid" on the claims chart.

The teacher realized the students were applying a definition of solids and liquids they learned in earlier grades and did not recognize that powders, such as the flour and baby powder, are made up of very small solid particles. She explained to me how she chose an analogy to develop the idea that solid matter can be in the form of tiny particles, and although it might seem to behave like a liquid, it has the properties of a solid. She chose an anchoring example—marbles in a jar. The students could see that the marbles filled the jar and could be poured out of the jar, but all agreed the marbles were a collection of solids.

She then used a bridging analogy to move from the anchoring example to something closer to the flour example. She filled a jar with salt. The students agreed that it seem to take the shape of the jar and could be poured out. The teacher then gave the students hand lenses to examine the tiny grains of salt. The students agreed they were "teeny tiny" bits of solids. They were so tiny they could not be seen as individual particles of solid when they filled the jar. The teacher then used this bridging analogy to have students rethink whether they still thought flour was a liquid. The students gave up their idea that flour is a liquid, and used their new model of a collection of tiny solid particles to describe flour as a solid.

—Page Keeley

When discussing conceptual change and changing mental models, the role of two other factors becomes evident: *epistemology* and *interest*. The truth is, we do not often think about the epistemology of our students, but it is increasingly important when discussing the use of models in science instruction. Students fall into roughly two categories (or on some part of the continuum between the two) in terms of what they believe about knowledge and how we come to know. There are those who believe that knowledge is truth—static and unchangeable—and those who view knowledge as being tentative, adaptive, and evolving. The researchers call the former *epistemologically naive* and the latter *epistemologically sophisticated* (Hofer and Pintrich 1997).

Studies have shown that students who are epistemologically naive are less likely to modify their conceptual models, even in the face of logic and evidence (Andre and Windshitl 2003). It makes sense that those students who see knowledge as certain and absolute would be more reluctant to make changes in models that have served them well, even when they are challenged. This kind of student would rather memorize facts, and let us know that they would just like us to "tell me the right answer." It also shows that students who see knowledge as open to revision have a different attitude toward conceptual change. But both types of students can hold two or more different conceptual viewpoints even if they are in conflict with each other.

It may be important for teachers themselves to look at how they view knowledge in their epistemological views since it will certainly affect their teaching. Teachers who have a more open epistemology are more likely to teach with a more constructivist pedagogy than those who believe in the static nature of knowledge.

Interest in the topic also has an impact on building and changing models and conceptions. As one might expect, the more a student has interest in a topic, the more effort will be produced toward learning. It could be a double-edged sword, however, since with interest may come experience and strong belief in a naive theory could be linked to stubbornness toward criticism and change. A study by Andre and colleagues (1999) showed that elementary school females were less interested in the physical sciences than males. So these girls might come to class with a less vested interest in their models about the physical world than their male peers.

### Refutational Text

*Refutational text* is an explanation of the concept or model that starts with a naive idea and then refutes that idea with the accepted scientific model. Studies have shown that students using refutational text enjoy the exercise and are more likely to alter their models and conceptions than those using standard text material written about straight content in a didactic manner (Alvermann and Hynd 1989).

A refutational piece about a preconception would state that conception and explain why someone might think that was true. Then the argument would be given to point out the weak points of the original point of view and offer evidence to the contrary. As an example, let us try to show how such a text might look:

> *Many people think that the phases of the Moon are caused by the shadow of the Earth on the Moon's surface. There is evidence to support their view since when a lunar eclipse occurs, the shadow of the Earth passes over the Moon and creates a crescent at the beginning and at the end of the cycle. However, how can we explain the gibbous phase of the Moon since no shadow from our globe could cast such a shadow on another globe like the Moon?*

The text would go on to elaborate on the explanation of the phases of the Moon by showing how the sunlight is reflected off of the Moon's surface into our eyes on Earth.

### Understanding the Concept of Models

While the *NGSS* practice is about the development and use of models, it is also important to develop the *concept* of a model. Before students construct or use models, teachers should elicit students' preconceptions about what models are and are not. The formative assessment probe "Is It a Model?" (Figure 6.1) lists several examples of using models that prompt students to think about what a model is (Keeley and Tugel 2009). In the second part of the probe students explain their thinking about models. If the concept of a *model* has never been addressed in the curriculum, students tend to choose the physical models and explain models as things that you make out of materials. They fail to recognize other types of models such as mental and mathematical models. This is a clear indication to teachers that students need more experiences with a variety of models.

Furthermore, many of the models students are asked to make as part of a science project end up being more like art projects than scientific models. Consider the cell models made of candy, cake, or other

**Figure 6.1. "Is It a Model?" Probe**

## Is It a Model?

Is it a Model?

Below are listed things that students might do in a science class. Check off the things that are examples of using a model.

____ **A** building a paper airplane
____ **B** making an analogy (for example, the heart is like a pump)
____ **C** observing a bird's behavior at a bird feeder
____ **D** developing a mathematical equation to solve a science problem
____ **E** making a plant cell out of household materials
____ **F** analyzing whale migration patterns with a computer program
____ **G** building and testing a bridge made of toothpicks
____ **H** drawing an electrical circuit
____ **I** forming a mental image of molecules in the liquid state
____ **J** demonstrating the day/night cycle with a globe and flashlight
____ **K** dissecting a cow's bone
____ **L** watching a computer simulation of a hurricane
____ **M** going on a field trip to the Grand Canyon
____ **N** graphing the speed of a car
____ **O** watching a live video of an active volcano
____ **P** making a replica of a human heart out of clay
____ **Q** looking at blood cells under a microscope

Explain your thinking. How did you decide whether something is a model?

*Source:* Keeley and Tugel 2009.

edible materials, and the Cheerios or M&Ms atomic models. Is the task more about choosing creative materials and building a good-looking model than using the model for explanatory or predictive purposes (the way scientists use models)? The seminal publication *Science for All Americans* describes models this way:

> *A model of something is a simplified imitation of it that we hope can help us understand it better. A model may be a device, a plan, a drawing, an equation, a computer program, or even just a mental image. Whether models are physical, mathematical, or conceptual, their value lies in suggesting how things either do work or might work. For example, once the heart has been likened to a pump to explain what it does, the inference may be made that the engineering principles used in designing pumps could be helpful in understanding heart disease. When a model does not mimic the phenomenon well, the nature of the discrepancy is a clue to how the model can be improved. Models may also mislead, however, suggesting characteristics that are not really shared with what is being modeled. Fire was long taken as a model of energy transformation in the sun, for example, but nothing in the sun turned out to be burning. (AAAS 1988, p. 168)*

*NGSS* Appendix F (Science and Engineering Practices in the *NGSS*) includes a matrix for examining the progression of what students are expected to do for the practice of Developing and Using Models at different grade spans. As you examine Table 6.2, notice how this practice overlaps with other practices.

### *Planning and Carrying Out Investigations*

*A Framework for K–12 Science Education* summarizes this science practice this way:

> *Scientific investigation may be conducted in the field or the laboratory. A major practice of scientists is planning and carrying out a systematic investigation, which requires the identification of what is to be recorded and, if applicable, what are to be treated as the dependent and independent variables (control of variables). Observations and data collected from such work are used to test existing theories and explanations or to revise and develop new ones. (NRC 2012, p. 50)*

Everyone who is familiar with teaching science knows the term "hands-on." As we mentioned before, this term was the watchword of decades past, and still today holds a magic that draws children and teachers alike. Teachers especially are drawn to activity-oriented professional development workshops. Activity is good and achieves motivation and interest in lessons, but activity by itself is not sufficient. As mentioned earlier, the learning comes in the discussion and argument *after* the activity—what many call the "minds-on" part. This means that learning involves conceptual modification. Thus the term that we advocate is *hands-on, minds-on.*

## Table 6.2. Practice 2: Developing and Using Models

**Practice 2: Developing and Using Models**

| Grades K–2 | Grades 3–5 | Grades 6–8 | Grades 9–12 |
|---|---|---|---|
| Modeling in K–2 builds on prior experiences and progresses to include using and developing models (i.e., diagram, drawing, physical replica, diorama, dramatization, or storyboard) that represent concrete events or design solutions.<br>• Distinguish between a model and the actual object, process, and/or events the model represents.<br>• Compare models to identify common features and differences.<br>• Develop and/or use a model to represent amounts, relationships, relative scales (bigger, smaller), and/or patterns in the natural and designed world(s).<br>• Develop a simple model based on evidence to represent a proposed object or tool. | Modeling in 3–5 builds on K–2 experiences and progresses to building and revising simple models and using models to represent events and design solutions.<br>• Identify limitations of models.<br>• Collaboratively develop and/or revise a model based on evidence that shows the relationships among variables for frequent and regular occurring events.<br>• Develop a model using an analogy, example, or abstract representation to describe a scientific principle or design solution.<br>• Develop and/or use models to describe and/or predict phenomena.<br>• Develop a diagram or simple physical prototype to convey a proposed object, tool, or process.<br>• Use a model to test cause and effect relationships or interactions concerning the functioning of a natural or designed system. | Modeling in 6–8 builds on K–5 experiences and progresses to developing, using, and revising models to describe, test, and predict more abstract phenomena and design systems.<br>• Evaluate limitations of a model for a proposed object or tool.<br>• Develop or modify a model—based on evidence—to match what happens if a variable or component of a system is changed.<br>• Use and/or develop a model of simple systems with uncertain and less predictable factors.<br>• Develop and/or revise a model to show the relationships among variables, including those that are not observable but predict observable phenomena.<br>• Develop and/or use a model to predict and/or describe phenomena.<br>• Develop a model to describe unobservable mechanisms.<br>• Develop and/or use a model to generate data to test ideas about phenomena in natural or designed systems, including those representing inputs and outputs and those at unobservable scales. | Modeling in 9–12 builds on K–8 experiences and progresses to using, synthesizing, and developing models to predict and show relationships among variables between systems and their components in the natural and designed world(s).<br>• Evaluate merits and limitations of two different models of the same proposed tool, process, mechanism, or system in order to select or revise a model that best fits the evidence or design criteria.<br>• Design a test of a model to ascertain its reliability.<br>• Develop, revise, and/or use a model based on evidence to illustrate and/or predict the relationships between systems or between components of a system.<br>• Develop and/or use multiple types of models to provide mechanistic accounts and/or predict phenomena, and move flexibly between model types based on merits and limitations.<br>• Develop a complex model that allows for manipulation and testing of a proposed process or system.<br>• Develop and/or use a model (including mathematical and computational) to generate data to support explanations, predict phenomena, analyze systems, and/or solve problems. |

*Source:* Appendix F: Science and Engineering Practices in the *NGSS*. NGSS Lead States 2013, p. 53.

Students like to carry out investigations, especially if the interest is high. If we choose topics that segue into engineering projects, the motivation carries over and the engineering project has even more appeal. For example, we have been linking three resources from our books in workshops lately: a probe from the *Uncovering Student Ideas in Physical Science* series, "The Swinging Pendulum" (Figure 6.2, p. 110; Keeley and Harrington 2010), to determine students' preconceptions about what affects the period of a pendulum, followed by the story "Grandfather's Clock" from *Everyday Science Mysteries* (Konicek-Moran 2008) that allows the students to learn about identifying and controlling variables to find out which variable(s) affect the amount of time for the pendulum to swing forth and back. Then they are given the engineering problem mystery story "The Crooked Swing" (Konicek-Moran 2011) and use the knowledge gained in the pendulum investigation to solve a mystery about why a porch swing hanging from a tree branch won't swing straight and design a solution to fix the problem.

Research has suggested that children at all ages have trouble defining relevant variables and learning how to design investigations to control variables (Kuhn and Phelps 1982;

## Figure 6.2. "The Swinging Pendulum" Probe

**The Swinging Pendulum**

Gusti made a pendulum by tying a string to a small bob. He pulled the bob back and counted the number of swings the pendulum made in 30 seconds. He wondered what he could do to increase the number of swings made by the pendulum. If Gusti can change only one thing to make the pendulum swing more times in 30 seconds, what should he do? Circle what you think will make the pendulum swing more times.

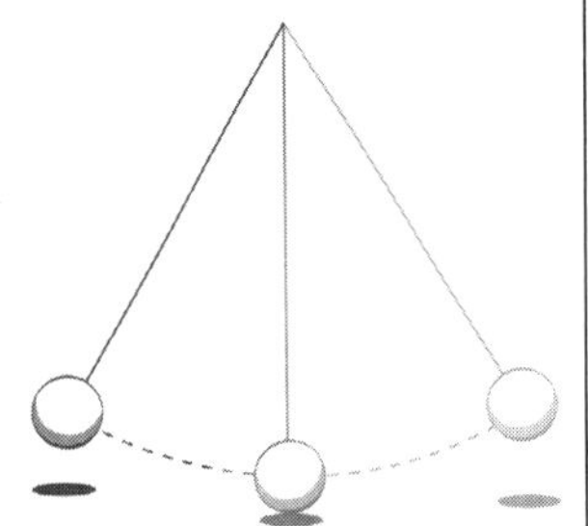

**A** Lengthen the string.

**B** Shorten the string.

**C** Change to a heavier bob.

**D** Change to a lighter bob.

**E** Pull the bob back farther.

**F** Don't pull the bob back as far.

**G** None of the above. All pendulums swing the same number of times.

Explain your thinking. What rule or reasoning did you use to select your answer?

*Source:* Keeley and Harrington 2010.

Schauble et al. 1995). Data gathering may also be a problem, since neophytes do not see the value of organizing data for future use. Helen Buttemer's article "Inquiry on Board" (2006) in *Science and Children* is a wonderful resource for teachers, offering a vivid example of using inquiry boards to kinesthetically isolate variables with moveable sticky notes and use them to design and conduct an investigation that involves variables. Inquiry boards were adapted from an approach used in England with primary students (Ward, Roden, Hewlett, and Foreman 2005). Buttemer explains, "This process of conducting an investigation—from framing the testable question to answering the question with evidence—requires the student to understand the anatomy of an investigation and the relationship between variables. We have found no better tool for introducing the logic of this relationship than inquiry boards" (2006, p. 35). Students can go through the process of identifying and controlling variables but still lack a conceptual understanding of why it is necessary in an experiment.

### Author Vignette

I had the opportunity to work with fourth graders whose teachers were not happy with the way they were using the unit on Mystery Powders from the Elementary Science Study (ESS) program. In this activity, students are presented with methods of testing for various household materials such as flour, baking powder, cornstarch, salt, sugar, and plaster of Paris. They try all of these tests and record them on a data sheet for reference. Then they are given a mixture made of several of these white powders and asked to perform the tests on the mixture to determine its contents. The teachers in the fourth-grade team felt the students were just going through the motions and not putting the knowledge gained into a useful and relevant pattern.

The team decided they needed to make the exercise more relevant to the students by writing a mystery story in which the tests and results might have more meaning. One of the teachers volunteered to write it. Her story related a series of mistakes made as she was preparing chocolate chip cookies, ending with three identical bowls, each containing some but not all of the necessary ingredients. The problem was that one of the bowls had all of the correct ingredients and the other two were contaminated with plaster of Paris due to workmen who knocked over some plaster of Paris while working in her kitchen while she was momentarily called away by an interruption. The students had to find out which bowl was the right one, and then they could claim the reward of their teacher baking cookies for the class to enjoy.

Alesia Peck, the writer of the story, gathered her class together and asked them how they proposed to find out the answer to the puzzle (and get the reward) and was amazed at how well they cooperated to design an investigation. As they discussed the problem together, they recalled their previous experiences with baking soda and vinegar, salt, sugar, and plaster of Paris. They only lacked the test for starch using iodine, which Alesia gave them. They decided to break into smaller groups and do the tests on samples taken from each bowl. Many stopped when they found the right bowl, but one girl, clearly a future scientist, insisted on testing the contents of all bowls for all ingredients, "just in case Ms. Peck forgot to add something" (Peck and Konicek 1998).

—Dick Konicek-Moran

The classic scientific method, which many of us remember from our own K–12 science education, is still taught today in some classrooms. For students to conceptually understand the scientific practice of planning and carrying out investigations, we must help them see that this is only one of the many ways that scientists plan and carry out investigations to develop scientific knowledge. As we have mentioned before, perhaps it would be better to refer to *a* scientific method rather than *the* scientific method, or to eliminate that term entirely, since it often implies a definite, linearly ordered series of steps all scientists follow.

*NGSS* Appendix F (Science and Engineering Practices in the *NGSS*) includes a matrix of the progression of what students are expected to do for the practice of Planning and Carrying Out Investigations at different grade spans. As you examine Table 6.3, notice how this practice overlaps with other practices.

### *Analyzing and Interpreting Data*

*A Framework for K–12 Science Education* says of the practice of Analyzing and Interpreting Data:

> *Scientific investigations produce data that must be analyzed in order to derive meaning. Because data usually do not speak for themselves, scientists use a range of tools—including tabulation, graphical interpretation, visualization, and statistical analysis—to identify the significant features and patterns in the data. Sources of error are identified and the degree of certainty calculated. Modern technology makes the collection of large data sets much easier, thus providing many secondary sources for analysis. (NRC 2012, p. 51)*

#### Data Collection

As we mentioned in Chapter 1, data do not come with an inherent structure. Structure must be imposed on data, say the authors *of Ready, Set, SCIENCE!* They mean that data can used for many purposes, but they must be organized to answer questions. Recording and accessing data efficiently is one of the most important lessons students can learn (Michaels, Shouse, and Schweingruber 2008).

## Table 6.3. Practice 3: Planning and Carrying Out Investigations

**Practice 3: Planning and Carrying Out Investigations**

| Grades K–2 | Grades 3–5 | Grades 6–8 | Grades 9–12 |
|---|---|---|---|
| Planning and carrying out investigations to answer questions or test solutions to problems in K–2 builds on prior experiences and progresses to simple investigations, based on fair tests, which provide data to support explanations or design solutions.<br>• With guidance, plan and conduct an investigation in collaboration with peers (for K).<br>• Plan and conduct an investigation collaboratively to produce data to serve as the basis for evidence to answer a question.<br>• Evaluate different ways of observing and/or measuring a phenomenon to determine which way can answer a question.<br>• Make observations (firsthand or from media) and/or measurements to collect data that can be used to make comparisons.<br>• Make observations (firsthand or from media) and/or measurements of a proposed object, tool, or solution to determine if it solves a problem or meets a goal.<br>• Make predictions based on prior experiences. | Planning and carrying out investigations to answer questions or test solutions to problems in 3–5 builds on K–2 experiences and progresses to include investigations that control variables and provide evidence to support explanations or design solutions.<br>• Plan and conduct an investigation collaboratively to produce data to serve as the basis for evidence, using fair tests in which variables are controlled and the number of trials is considered.<br>• Evaluate appropriate methods and/or tools for collecting data.<br>• Make observations and/or measurements to produce data to serve as the basis for evidence for an explanation of a phenomenon or to test a design solution.<br>• Make predictions about what would happen if a variable changes.<br>• Test two different models of the same proposed object, tool, or process to determine which better meets criteria for success. | Planning and carrying out investigations in 6–8 builds on K–5 experiences and progresses to include investigations that use multiple variables and provide evidence to support explanations or solutions.<br>• Plan an investigation individually and collaboratively, and in the design identify independent and dependent variables and controls, what tools are needed to do the gathering, how measurements will be recorded, and how many data are needed to support a claim.<br>• Conduct an investigation and/or evaluate and/or revise the experimental design to produce data to serve as the basis for evidence that meet the goals of the investigation.<br>• Evaluate the accuracy of various methods for collecting data.<br>• Collect data to produce data to serve as the basis for evidence to answer scientific questions or to test design solutions under a range of conditions.<br>• Collect data about the performance of a proposed object, tool, process, or system under a range of conditions. | Planning and carrying out investigations in 9–12 builds on K–8 experiences and progresses to include investigations that provide evidence for and test conceptual, mathematical, physical, and empirical models.<br>• Plan an investigation or test a design individually and collaboratively to produce data to serve as the basis for evidence as part of building and revising models, supporting explanations for phenomena, or testing solutions to problems. Consider possible confounding variables or effects and evaluate the investigation's design to ensure variables are controlled.<br>• Plan and conduct an investigation individually and collaboratively to produce data to serve as the basis for evidence, and in the design decide on types, how much, and accuracy of data needed to produce reliable measurements and consider limitations on the precision of the data (e.g., number of trials, cost, risk, time), and refine the design accordingly.<br>• Plan and conduct an investigation or test a design solution in a safe and ethical manner, including considerations of environmental, social, and personal impacts.<br>• Select appropriate tools to collect, record, analyze, and evaluate data.<br>• Make directional hypotheses that specify what happens to a dependent variable when an independent variable is manipulated.<br>• Manipulate variables and collect data about a complex model of a proposed process or system to identify failure points or improve performance relative to criteria for success or other variables. |

*Source:* Appendix F: Science and Engineering Practices in the *NGSS*. NGSS Lead States 2013, p. 55.

## Author Vignette

The "Mystery Powder" activity caused problems for the students in the previous vignette because they were not ready to see the importance of collecting data in a systematic way. When Alesia, the teacher, asked them how they intended to keep track of their data, they said they would figure it out as they worked. Alesia did not press the issue, since they were doing such a great job designing the investigation. She might have insisted that they record their data on prepared sheets but she decided that allowing them to record their data in a helter-skelter manner would allow them to see for themselves the importance of planning ahead to organize and record data. The students did indeed find this out, when some of the groups tried to remember which set of data belonged to which bowl. In subsequent investigations, data and data organization became very important to them (Peck and Konicek 1998). As Eleanor Duckworth stated, "The problem is that they don't see the problem" (2012).

—Dick Konicek-Moran

### Extrapolation and Prediction

Interpreting data is important too. A class of second graders had carefully designed a graph showing how the length of the tree shadow caused by the Sun changed over time, a yearlong project. As the end of school approached in June, the graphic lengths of the shadows were getting shorter and shorter as it approached the summer solstice. The children extrapolated the trend on the graph and predicted that by mid July or August, there would be no shadow at all! There had been no mention of the cyclical nature of the phenomenon in their instruction before, so who can blame them? Luckily, the third-grade team agreed to continue the data gathering the following September so the students could see the seasonal cycle.

*Extrapolation* or *prediction* from data must be embedded in a well-planned instructional strategy, with an emphasis on interpreting data as well as collecting it. A middle school teacher related a story about students in his class who were recording temperatures plotted against time as heated water cooled. The students predicted that at a given time the temperature would be well below room temperature, in fact freezing by that night. The concept of equilibrium never occurred to them! A golden opportunity to bring up a new concept thereby presented itself.

Beyond the problems associated with extrapolation lie the issues surrounding the interpretation of data and the uses of such tools as graphs, tables, and statistical representations. It is important for teachers to understand what sorts of visual representations of data are going to be effective for different age groups. For example, we have found that in counting or weighing activities, centigram cubes or beads that stick together are useful as data gatherers for young children. They become three-dimensional graphs for younger children and can be compared in a nonmathematical way as well as numerically.

Graphing data and analyzing it points out the disparity in procedural tasks versus tasks that involve conceptual understanding. For example, sometimes the procedure of constructing a graph is overemphasized at the expense of analyzing a graph. Including a title, labeling the *x* and *y* axes, determining scale, and correctly plotting points may be the focus of that task or even the assessment. However, when students are asked to interpret a graph, they often interpret it literally. Middle and even high school students who are asked to interpret distance/time or position/time graphs involving steep, gradual, and zero slope will often explain the data as going up or down a hill, or traveling on a flat surface. It is important for students to conceptually understand what a graph represents and to be able to tell the story of the graph, not just construct one.

## Author Vignette

I once had a discussion during a workshop with a mathematics colleague about similar conceptual difficulties both mathematics and science students have in interpreting graphs. My math colleague told me how she would often use the interesting variety of graphs included in the front inset of the newspaper *USA Today* to help her middle school mathematics students use the practice of interpreting data from graphs. One issue included a graph titled "7-Up Loses Its Fizz." She asked her students to do a quick write to explain what the graph showed. Her students wrote that the graph showed how 7-Up loses its fizz the longer an opened can of 7-Up sits out on a table. The actual graph depicted how sales of 7-Up have decreased over the years! The students did not even look at the relationship between the variables—they interpreted the title in their own way without even considering what the data actually showed!

—Page Keeley

### Observation and Inference

Let us imagine that we come upon a small animal. We make observations about its anatomy and its behavior. We try to explain it by making an *inference*. Students might say, for example, "The worm moves along the edge of the box because it is scared," or, "The worm doesn't like the light." An inference may or may not be right—it must be tested in a controlled experiment. Neither one of these inferences can be tested, and so are not scientifically valid. Students learn to make inferences as a reading strategy when using text; however, making inferences in science is different from making inferences while reading a novel.

> ## Author Vignette
>
> A friend of mine, Leonard Amburgy, a naturalist who worked in a nearby nature center, introduced me to a game called "What Does It Tell You and What Do You Want to Know?" In this exercise, students who visit the nature center are presented with a mystery organism, and the organism "tells" you only what you are able to observe—no explanations and if possible the animal itself remains anonymous. We often used mealworms (*Tenebrio molitor*). Students may observe that, "the organism moves straight ahead at an average velocity of 25 mm per minute," or "the organism has 25 segments." Later on when the observations have run out, we go to the second part, "What Do You Want to Know?" In this half of the activity, we design investigations to find out something about the organism that it did not "tell" us. From there we can make a *conclusion* such as, "The organism only eats bran when presented with five different kinds of food."
>
> —Dick Konicek-Moran

### Inferential Evidence and Inferential Distance

NSTA's middle school journal *Science Scope* has an excellent article about evaluating the strength of evidence when interpreting data. The article, "Helping Students Evaluate the Strength of Evidence in Scientific Arguments," mentions the terms *inferential evidence* and *inferential distance*. The authors point out how the number of observations or trials in an experiment or the validity of information gained from another source (*inferential evidence*) can either shorten or lengthen the distance between the observation and the strength of the claim (*inferential distance*). The shorter the distance between observation and claim, the stronger the claim. They identify three factors that *lengthen* the inferential distance:

(1) basing an inference about a large population on a small number of observations or on a limited number of subjects; (2) basing an inference about a large or different model on a smaller model, for example, making an inference on a pond or lake based on observations made in an aquarium; and (3) failing to look for alternative explanations. Using this model for evaluating claims and inferences can be very useful for students and teachers as they critique the strengths of arguments (Brodsky, Falk, and Beals 2013).

### Validity of Data and Data Sources

Being a responsible citizen requires that we be responsible for ascertaining the validity of data. We need to know how to read and interpret a graph and to find sources of information that are reliable. Do we believe Fox News, MSNBC, NPR, Wikipedia, or the journal *Science*? How do we deal with the studies whose findings on diet change from week to week? How do we decide on the best buy in the products we purchase? What can we do to find out if the human carbon footprint is really responsible for the changes in the Earth's climate? With our analyses of data, we vote in referendums and elections and are in some way a part of decisions that affect everyone. Therefore, the practice of Analyzing and Interpreting Data is essential not just for students in science class or for those doing scientific work, but for all citizens of this world who make decisions that affect their lives, their environment, and the welfare of society.

One of the best resources we recommend for helping students (and teachers) understand how to interpret data is an NSTA Press book, *The Basics of Data Literacy: Helping Your Students (and You!) Make Sense of Data* (Bowen and Bartley 2013). This book addresses types of variables and data, ways to structure and interpret data, basic statistics, and survey data. Because data are so central to developing conceptual understanding of scientific ideas, including the core ideas in the *NGSS*, the ability to understand and work with data is a practice that needs to be continuously honed for both students and teachers.

*NGSS* Appendix F (Science and Engineering Practices in the *NGSS*) includes a matrix of the progression of what students are expected to do for the practice of Analyzing and Interpreting Data. As you examine Table 6.4 (p. 118), notice how this practice overlaps with other practices.

## *Using Mathematics and Computational Thinking*

*A Framework for K–12 Science Education* summarizes this practice:

> *In science, mathematics and computation are fundamental tools for representing physical variables and their relationships. They are used for a range of tasks, such as constructing simulations, statistically analyzing data, and recognizing, expressing, and applying quantitative relationships. Mathematical and computational approaches enable predictions of the behavior of physical systems, along with the testing of such predictions. Moreover, statistical techniques are invaluable for assessing the significance of patterns or correlations. (NRC 2012, p. 51)*

## Table 6.4. Practice 4: Analyzing and Interpreting Data

**Practice 4: Analyzing and Interpreting Data**

| Grades K–2 | Grades 3–5 | Grades 6–8 | Grades 9–12 |
|---|---|---|---|
| Analyzing data in K–2 builds on prior experiences and progresses to collecting, recording, and sharing observations.<br>• Record information (observations, thoughts, and ideas).<br>• Use and share pictures, drawings, and/or writings of observations.<br>• Use observations (firsthand or from media) to describe patterns and/or relationships in the natural and designed world(s) in order to answer scientific questions and solve problems.<br>• Compare predictions (based on prior experiences) to what occurred (observable events).<br>• Analyze data from tests of an object or tool to determine if it works as intended. | Analyzing data in 3–5 builds on K–2 experiences and progresses to introducing quantitative approaches to collecting data and conducting multiple trials of qualitative observations. When possible and feasible, digital tools should be used.<br>• Represent data in tables and/or various graphical displays (bar graphs, pictographs, and/or pie charts) to reveal patterns that indicate relationships.<br>• Analyze and interpret data to make sense of phenomena, using logical reasoning, mathematics, and/or computation.<br>• Compare and contrast data collected by different groups in order to discuss similarities and differences in their findings.<br>• Analyze data to refine a problem statement or the design of a proposed object, tool, or process.<br>• Use data to evaluate and refine design solutions. | Analyzing data in 6–8 builds on K–5 experiences and progresses to extending quantitative analysis to investigations, distinguishing between correlation and causation, and basic statistical techniques of data and error analysis.<br>• Construct, analyze, and/or interpret graphical displays of data and/or large data sets to identify linear and non-linear relationships.<br>• Use graphical displays (e.g., maps, charts, graphs, and/or tables) of large data sets to identify temporal and spatial relationships.<br>• Distinguish between causal and correlational relationships in data.<br>• Analyze and interpret data to provide evidence for phenomena.<br>• Apply concepts of statistics and probability (including mean, median, mode, and variability) to analyze and characterize data, using digital tools when feasible.<br>• Consider limitations of data analysis (e.g., measurement error) and/or seek to improve precision and accuracy of data with better technological tools and methods (e.g., multiple trials).<br>• Analyze and interpret data to determine similarities and differences in findings.<br>• Analyze data to define an optimal operational range for a proposed object, tool, process, or system that best meets criteria for success. | Analyzing data in 9–12 builds on K–8 experiences and progresses to introducing more detailed statistical analysis, the comparison of data sets for consistency, and the use of models to generate and analyze data.<br>• Analyze data using tools, technologies, and/or models (e.g., computational, mathematical) in order to make valid and reliable scientific claims or determine an optimal design solution.<br>• Apply concepts of statistics and probability (including determining function fits to data, slope, intercept, and correlation coefficient for linear fits) to scientific and engineering questions and problems, using digital tools when feasible.<br>• Consider limitations of data analysis (e.g., measurement error, sample selection) when analyzing and interpreting data.<br>• Compare and contrast various types of data sets (e.g., self-generated, archival) to examine consistency of measurements and observations.<br>• Evaluate the impact of new data on a working explanation and/or model of a proposed process or system.<br>• Analyze data to identify design features or characteristics of the components of a proposed process or system to optimize it relative to criteria for success. |

*Source:* Appendix F: Science and Engineering Practices in the *NGSS*. NGSS Lead States 2013, p. 57.

Trying to accommodate data into your thinking processes without using mathematics is a difficult task. Instructionally speaking, you can't have one without the other. After all, mathematics is frequently referred to as the language of science.

Back in 1978, a brilliant educator and a man ahead of his time, David Hawkins, wrote an essay modified from a talk he had given earlier at the Massachusetts Institute of Technology. He called it "Critical Barriers to Science Learning" (1978). Hawkins identified *size* and *scale* as two extremely important concepts that, if not mastered, would stand in the way of understanding a myriad of other concepts. In this, he was in the forefront of suggesting that science instruction be organized around key concepts, or as we know them today, *crosscutting concepts*. Size and scale are inevitably tied up with numbers.

Science and mathematics have gone hand-in-hand since the enlightenment years and possibly before. Without quantifying his observations, Galileo would never have made his discoveries on gravity and acceleration. Galileo did not have the luxury that we have today in the use of computer technology, and so his calculations had to be done the slow and careful way. Tycho Brahe (1546–1601), the Danish astronomer, was a remarkable observer of the "heavenly bodies," and it was his quantification of the motions of the planets that provided the precise data used by Kepler, Newton, and others to construct the laws of planetary motion.

By using mathematics, we mean more than just arithmetic and algorithms. We mean using geometry and at least elementary statistics, graphing, and modeling tactics as well (for example, Figure 6.3). Students need to understand *ratio,* which allows us to deal with two properties of matter at the same time (e.g., volume and mass when considering density) or in physics to understand the relationships between force, mass, and acceleration. Understanding probability is essential to genetics and medical sciences. Knowledge of simple algebraic processes and functions serve students well in science, as well as for the rest of their lives.

**Figure 6.3. Example of a Time-Distance Graph**

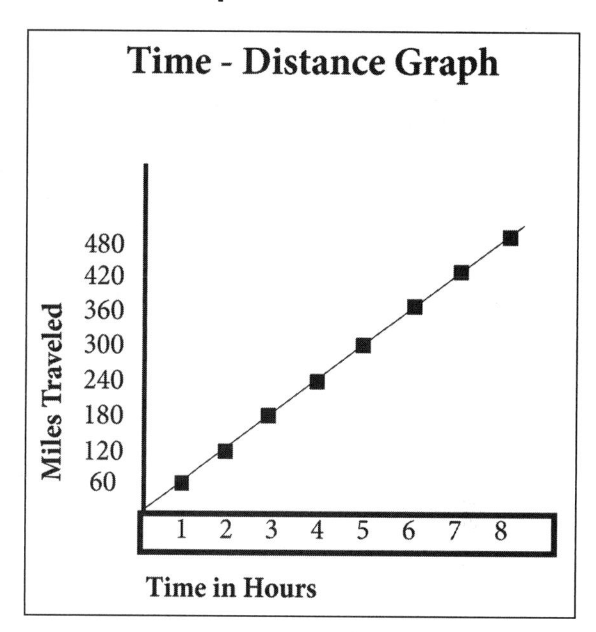


And of course, we now have the wonder of computer technology, and as we write this book, we often ask ourselves, "What would we do without the internet?" It has allowed us to access scientific and educational journals, obtain information on just about any topic, do calculations that would take hours if not more to do by hand, graph data, record differences in temperature, and scads of other things. The personal computer has changed everyone's social as well as professional life, and of course provides us with interactive games and simulations for model testing. It is now estimated (without statistical evidence) that people spend more time on their mobile devices than they do on their home tools. It is a common joke that parents and grandparents ask their children for advice on how to use technology. We are teaching a generation that has grown up with technology we never dreamed of and it is incumbent upon us to make use of this knowledge possessed by our students.

### Statistical Thinking

Students need to understand what *averaging* means and its limitations in the interpretation of data. Data often include numbers that represent "apples and oranges," and therefore a simple average will end up in a mess. These apples have to be sorted out from the oranges and mathematized separately. But rather than memorize definitions of traditional statistical markers (such as the *mean, median,* and *mode*), students need multiple opportunities to work with data and use these statistical concepts. Furthermore, instruction needs to move beyond procedural understanding (e.g., being able to calculate the mean) and include a focus on conceptual understanding by connecting the mathematics to the scientific content (e.g., what does the mean tell us about this phenomenon?). Students can use mathematical procedures without any connection to the scientific ideas as evidenced in the following vignette.

## Author Vignette

During a professional development session targeting middle school students' energy-related ideas, I asked teachers to bring samples of student work from the "Mixing Water" probe (Keeley, Eberle, and Tugel 2007). The probe presented students with two glasses half-filled with equal volumes of water. The water in one glass was 50°C. The water in the other glass was 10°C. One glass was poured into the other, stirred, and the temperature of the resulting mixture was measured. Students were asked to predict what they thought the temperature of the combined full glass of water would be. They were given the choices of 20°C, 30°C, 40°C, 50°C, and 60°C. As the teachers sorted the student responses, they were surprised to see a large number of students add the temperatures together for a resulting temperature of 60°C. They were equally surprised to see a number of students subtract the cooler temperature from the warmer temperature for a resulting temperature of 40°C (at least this acknowledges that the final temperature lies somewhere in between). The students basically used subtractive or additive reasoning without a conceptual understanding of *qualitative temperature*. How could "warm" and "cool" be added together to get something even "warmer"? In other words, they used mathematics without considering the temperatures of the two volumes or the scientific idea of what was happening when warm water comes in contact with cooler water. The teachers were even more surprised when they examined the reasoning of the students who selected the best answer, 30°C. The students reasoned that the temperature would be the average of the two temperatures, but they were unable to offer an explanation of why. They used mathematics but couldn't connect it to what was happening when the two volumes mixed. The teachers concluded that their students needed to develop "temperature sense," starting with qualitative temperature tasks before presenting them with quantitative

> temperature tasks, and that students could use mathematical ideas devoid of scientific reasoning. They then brainstormed ways to use a model to show what is happening when a volume of warm water comes in contact with an equal volume of cooler water.
>
> —Page Keeley

### Understanding Visual Representations of Numbers (Graphs)

Students need to understand how to read graphs, and how to choose the type of graphs that will allow them to understand their data and present these data to others. Histograms, pie charts, and bell curve (or frequency distribution) charts all provide different, effective ways to visually represent numerical data. When graphing a bell-shaped curve chart, students need to look at the curve of their data and notice the distribution of data and understand the difference among *mean, median,* and *mode* and the limits of recorded numbers on the edges of the graph. They need to understand rate of change and slope. A graph can be interpreted by the slope of the lines, such as in the "Go-Cart Test Run" graph (Figure 6.4). A student who has conceptual understanding of slope can interpret the graph by the slope of the lines. A straight line with a steep slope indicates constant motion at a greater speed than a straight line with a more gradual slope. When the line on a graph is horizontal, that means that as time goes by, the position of an object is not changing. The horizontal line on the "Go-Cart Test Run" graph means that the go-cart stops at the end of the motion.

**Figure 6.4. "Go-Cart Test Run" Probe**

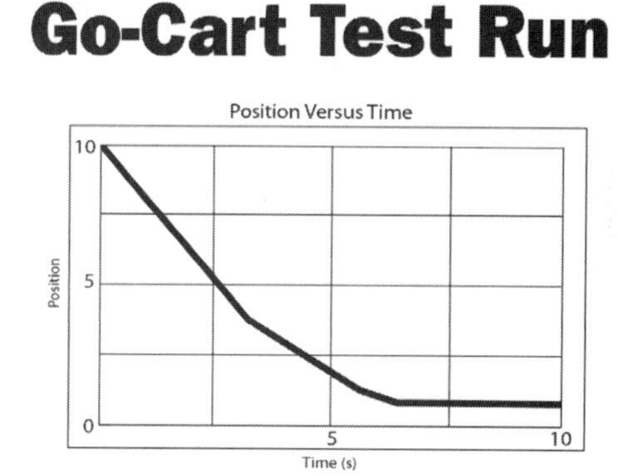


Jim and Karen have built a go-cart. They take their go-cart for a test run and graph its motion. Their graph is shown above. They show the graph to their friends. This is what their friends say:

**Bill:** "Wow, that was a steep hill! You must have been going very fast at the bottom."

**Patti:** "I think you were going fast at first, but then you slowed down at the end."

**Kari:** "I think you must have hit something along the way and come to a full stop."

**Mort:** "It looks like you were going downhill and then the road flattened out."

Circle the name of the friend you think best describes the motion of the go-cart, based on the graph. Explain why you agree with that friend.

*Source:* Keeley and Harrington 2012.

## Author Vignette

Students often will focus on the mode in data collected because it identifies the number most often collected. As an example, while working with a group of second graders, we offered each of them three snow-pea pods from the same batch picked from a vegetable stand. They counted the number of peas in each pod and placed markers on a paper that compared number of peas in each pod against the number of pods. The graph (a histogram) looked like a nicely bell-shaped curve with an obvious central tendency, the mode being 7.

We reserved a few pods and left them in a bag and after reviewing the data on the graph, asked the students to use those data to predict how many peas they thought might be in each pod randomly picked from the bag. Most students predicted with utmost certainty, the number 7. Others refused to pick one number and instead chose a *range*, from 5 to 8. Yet others picked random numbers because those numbers were "their lucky numbers." In other words, the majority of the students picked the mode or most often repeated number, while some students realized that the graph certainly showed a probability but also other possibilities. Some also suggested that there might be much higher or lower numbers of peas in the pods because the remaining pods might fill in the gap created by the low and high numbers because there were so few representations on the graph at the extremes. When we did pick the rest of the pods from the bag it turned out that the children who chose the range were much more successful than those that chose a single number, the mode. Remember that next time you place a bet at a roulette wheel.

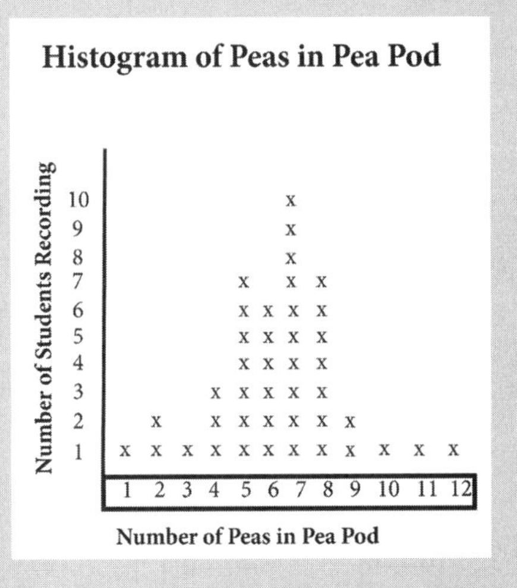


—Dick Konicek-Moran

## Understanding Concepts First

Paul Hewitt has written a book for high school entitled *Conceptual Physics,* now in its 12th edition. He tackled all of the areas of physics with analogies and connections to everyday happenings in students' lives. It was an immediate success because it promoted the conceptual understanding of physical phenomena, while advocating the later use of the mathematics involved—quite a different approach from the traditional way physics had been taught. Students found that they were more likely to understand the mathematics involved in the study of physical science if they had a conceptual view first (Hewitt 2006). *Active Physics,* developed by NSTA past-president Dr. Arthur Eisenkraft, takes a similar approach so high school freshman can develop a conceptual understanding of physics prior to learning and using advanced mathematics in later grades.

**Figure 6.5. "How Far Did It Go?" Probe**

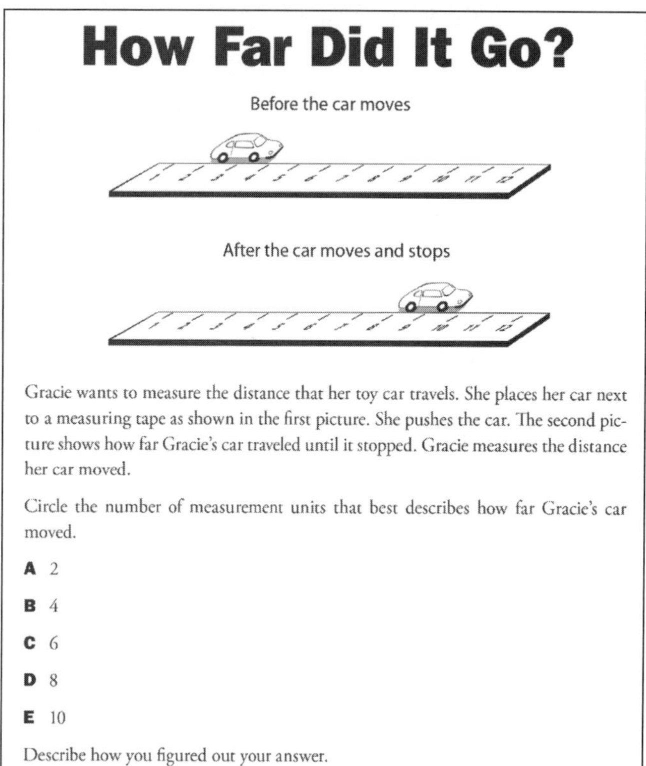


*Source:* Keeley and Harrington 2010.

The stories in the *Everyday Science Mysteries* series (Konicek-Moran 2008–2013) often create a common problem that can be used to get all students on board and more receptive to using numerical data. The open-ended stories give them conceptual ideas to work with before designing and carrying out experiments and collecting data and working with the mathematics involved.

Measurement is used extensively in science. Students become familiar with measurement instruments such as tape measures, balance scales, graduated cylinders, thermometers, and more. Yet while they can read and record the measurement on a measurement scale, students of all ages may do this without a conceptual understanding of *measurement.* Often the emphasis has been on the skill or the procedure of measuring and not the concept of measurement. For example, the assessment probe "How Far Did It Go?" (Figure 6.5) reveals whether students take into account the non-zero origin on a measurement scale or if they merely read the number on the endpoint. Mathematical studies reveal that few children recognize that any point on a measurement scale can serve as a starting point. Even students up through fifth grade have been shown to lack this knowledge (Lindquist and Kouba 1989). Furthermore, students are inconsistent in their approach to measurement. Some students will start at the back tires and end with the front tires or vice versa. Measurement is a mathematical concept that needs to be conceptually understood, not just procedurally taught.

*NGSS* Appendix F (Science and Engineering Practices in the *NGSS*) includes a matrix of the progression of what students are expected to do for the practice of Using Mathematics and Computational Thinking at different grade spans. As you examine Table 6.5, notice how this practice overlaps with other practices.

## Table 6.5. Practice 5: Using Mathematics and Computational Thinking

**Practice 5: Using Mathematics and Computational Thinking**

| Grades K–2 | Grades 3–5 | Grades 6–8 | Grades 9–12 |
|---|---|---|---|
| Mathematical and computational thinking in K–2 builds on prior experience and progresses to recognizing that mathematics can be used to describe the natural and designed world(s).<br>• Decide when to use qualitative vs. quantitative data.<br>• Use counting and numbers to identify and describe patterns in the natural and designed world(s).<br>• Describe, measure, and/or compare quantitative attributes of different objects and display the data using simple graphs.<br>• Use quantitative data to compare two alternative solutions to a problem. | Mathematical and computational thinking in 3–5 builds on K–2 experiences and progresses to extending quantitative measurements to a variety of physical properties and using computation and mathematics to analyze data and compare alternative design solutions.<br>• Decide if qualitative or quantitative data are best to determine whether a proposed object or tool meets criteria for success.<br>• Organize simple data sets to reveal patterns that suggest relationships.<br>• Describe, measure, estimate, and/or graph quantities (e.g., area, volume, weight, time) to address scientific and engineering questions and problems.<br>• Create and/or use graphs and/or charts generated from simple algorithms to compare alternative solutions to an engineering problem. | Mathematical and computational thinking in 6–8 builds on K–5 experiences and progresses to identifying patterns in large data sets and using mathematical concepts to support explanations and arguments.<br>• Use digital tools (e.g., computers) to analyze very large data sets for patterns and trends.<br>• Use mathematical representations to describe and/or support scientific conclusions and design solutions.<br>• Create algorithms (a series of ordered steps) to solve a problem.<br>• Apply mathematical concepts and/or processes (e.g., ratio, rate, percent, basic operations, simple algebra) to scientific and engineering questions and problems.<br>• Use digital tools and/or mathematical concepts and arguments to test and compare proposed solutions to an engineering design problem. | Mathematical and computational thinking in 9–12 builds on K–8 experiences and progresses to using algebraic thinking and analysis, a range of linear and non-linear functions, including trigonometric functions, exponentials and logarithms, and computational tools for statistical analysis to analyze, represent, and model data. Simple computational simulations are created and used based on mathematical models of basic assumptions.<br>• Create and/or revise a computational model or simulation of a phenomenon, designed device, process, or system.<br>• Use mathematical, computational, and/or algorithmic representations of phenomena or design solutions to describe and/or support claims and/or explanations.<br>• Apply techniques of algebra and functions to represent and solve scientific and engineering problems.<br>• Use simple limit cases to test mathematical expressions, computer programs, algorithms, or simulations of a process or system to see if a model "makes sense" by comparing the outcomes with what is known about the real world.<br>• Apply ratios, rates, percentages, and unit conversions in the context of complicated measurement problems involving quantities with derived or compound units (such as mg/mL, $kg/m^3$, acre-feet, etc.). |

*Source:* Appendix F: Science and Engineering Practices in the *NGSS*. NGSS Lead States 2013, p. 59.

### *Constructing Explanations and Designing Solutions*

*A Framework for K–12 Science Education* summarizes this science practice:

> *The goal of science is the construction of theories that can provide explanatory accounts of features of the world. A theory becomes accepted when it has been shown to be superior to other explanations in the breadth of phenomena it accounts for and in its explanatory coherence and parsimony. Scientific explanations are explicit applications of theory to a specific situation or phenomenon, perhaps with the intermediary of a theory-based model for the system under study. The goal for students is to construct logically coherent explanations of phenomena that incorporate*

> *their current understanding of science, or a model that represents it, and are consistent with the available evidence. (NRC 2012, p. 52)*

As mentioned before, students have been constructing explanations all their lives, and they bring these with them to school. We also know that these explanations or naive theories do not come completely thought out and tested:

> *Scientific knowledge in many domains ... consist(s) of formally specified entities and the relationships posited as existing between them. The point is that, even in relatively simple domains of science, the concepts used to describe and model the domain are not revealed in an obvious way by reading the "book of nature." Rather they are constructs that have been invented and imposed on phenomena in attempts to interpret and explain them, often as results of considerable intellectual struggles. (Driver et al. 1994, p. 60)*

Since the ideas in science and the practices therein are being used to initiate our students into the scientific community, our role is to help our students make personal sense of the practices and of the knowledge that is now accepted by the scientific community at large. Bonnie Shapiro, whose work we have examined at great length in Chapter 4, noted that we are asking students to join a world of accepted ideas and to put their own ideas aside (1994), which often can be painful. Our job as science teachers is to bring our students as close as possible to understand the accepted scientific concepts. We say "as close as possible" because we are well aware that some of the understandings will come in someone else's classroom, perhaps years from when they were with us. We believe that when helping students develop new explanations, they need to understand that a theory *explains* a phenomenon or an experimental result. Students need help finding patterns and applying these patterns to broader meanings. They also need help finding new questions in the data as well as using existing data to answer new questions.

One of the best tools in helping students conceptually understand what a scientific explanation is and how to construct it has been the development of the C-E-R Framework: Claims, Evidence, and Reasoning (McNeill and Krajcik 2008). A scientific explanation includes three parts:

- Claim: A statement that answers the question or problem students are investigating.
- Evidence: The data that supports the claim. Data can be qualitative or quantitative.
- Reasoning: The link between the evidence and the claim. Uses scientific concepts or principles to explain how the evidence supports the claim.

Put all three parts together and they make a scientific explanation. The C-E-R Framework has helped students develop a conceptual understanding of how a scientific explanation differs from an ordinary explanation. It helps scaffold the process of constructing an explanation, whether through writing or orally so students can meaningfully discuss their data and use scientific concepts and principles when engaged in discussion and argument. With continued practice and feedback using the C-E-R Framework, students improve in their ability to construct a scientific explanation.

*A Framework for K–12 Science Education* summarizes the engineering practice of designing solutions this way:

> *Engineering design, a systematic process for solving engineering problems, is based on scientific knowledge and models of the material world. Each proposed solution results from a process of balancing competing criteria of desired functions, technological feasibility, cost, safety, aesthetics, and compliance with legal requirements. There is usually no single best solution but rather a range of solutions. Which one is the optimal choice depends on the criteria used for making of evaluations. (NRC 2012, p. 52)*

When it comes to engineering, there are several ways to encourage students to understand the processes needed to engage in engineering. One is to present them with a faulty product and challenge them to improve it using what they know about science. One investigation we have used very successfully with both adults and students is the "Tin Can Telephone," from *Everyday Physical Science Mysteries* (Konicek-Moran 2013c). This story would fit into a category of product improvement activities. Two girls who live next door to each other want to have a private telephone. They are introduced to the tin can telephone and then, when they have trouble understanding each other, determine there are variables that can be changed to get better reception. They must identify the problem, try different models of the product, test them, revise them, decide what kind of data they can use to determine improvements and then, after evaluating the product, build the improved model and test it again. In the classroom, they would also have to communicate their solution to others.

You may have observed that sometimes students go overboard in identifying variables they think are important. They need to learn to identify those variables that have a causal relationship and distinguish those from the variables that are at best superficially related to an outcome. This is borne out in research done by Schauble and colleagues. They also found that students were more likely to "try it out and see" when working on an engineering task; and to work more from a theory-based approach when working on a science-related problem (Schauble, Klopfer, and Raghavan 1991). In our efforts, we find that engineering problems based on science concepts go very well together and allow students to "apply" new knowledge in a practical way and validate their understanding at the same time.

In what order should they be taught? One study found that their students worked better in designing science experiments if they did the engineering problem first (Sneider, Kurlich, Pulow, and Friedman 1984). This paralleled the findings of Schauble also. We, in our work, have found the opposite to be true as well—a newly learned science concept is reinforced if followed soon by an engineering task. Teacher experience will probably be the deciding factor as to which comes first, "the chicken or the egg."

## Author Vignette

It might be of interest to know that the Environmental Education (EE) section of the National Park Service is taking the *Next Generation Science Standards* seriously and working them into their curriculum. An EE park ranger who works at the Cape Cod National Seashore assured me that their new curriculum focuses much more on student inquiry than giving out facts. He related how on a field trip to a salt marsh on the Cape, students as young as 12 or 13 were asked to define problems related to the persistent rise of ocean water and to suggest ways in which they might solve the problem of rising water on the marsh and on the homes that have been grandfathered there for many years. As a former engineer, he was awed by the many responses given by the students and especially by the imaginative solutions they suggested.

—Dick Konicek

This section would not be complete without mentioning the NSTA Toshiba Exploravision K–12 student competition and the Invention Convention alternative to science fairs. Both encourage students to identify a problem and then to propose a solution or solve it through creative inventions. One technique used by the fair planners is the "What Bugs Me?" prompt, which allows students to uncover problems by way of a pet peeve. Example: One young student was irked by the fact that she could never reach the coat hangers to hang up her coat. She invented a long-shanked coat hanger for small people that would reach the bar and still be accessible to individuals who could not reach the bar.

## Author Vignette

While I was on a sabbatical in the Wimbledon area of the United Kingdom, I was asked on several occasions to judge engineering contests in local schools. Students were given tasks such as, "In one hour build a platform for rolling a marble that takes exactly one minute from release to finish." The problems were for the most part, engineering types of tasks. Students

> had no prior knowledge of the problem and worked as teams to build whatever was needed to solve the problem in the allotted time. Materials were provided. Students were so adept at doing so that we found it necessary to use stopwatches that were capable of measuring time in hundredths of a second to determine winners. These kinds of "fairs" were both engaging and instructive. Students learned to identify a problem, develop a model, try it, and evaluate it, as well as learn to work together as a group.
>
> —Dick Konicek-Moran

*NGSS* Appendix F (Science and Engineering Practices in the *NGSS*) includes a matrix of the progression of what students are expected to do for the practice of Constructing Explanations and Designing Solutions. As you examine Table 6.6, notice how this practice overlaps with other practices.

### *Engaging in Argument From Evidence*

*A Framework for K–12 Science Education* summarizes this science practice:

> *In science, reasoning and argument are essential for identifying the strengths and weaknesses of a line of reasoning and for finding the best explanation for a natural phenomenon. Scientists must defend explanations, formulate evidence based on a solid foundation of data, examine their own understanding in light of the evidence and comments offered by others, and collaborate with peers in searching for the best explanation for the phenomenon being investigated.* (NRC 2012, p. 52)

Recently, we have become more aware of the importance of children talking in classrooms and the importance of language. Lev Vygotsky pointed out in his book *Thought and Language* that when children use words, it helps them develop concepts (1962).

#### Claims-Evidence-Reasoning (C-E-R) Framework and Argumentation

The Claims-Evidence-Reasoning Framework for constructing scientific explanations goes hand-in-hand with engaging in argument with evidence. When students engage in scientific argumentation, their different explanations are compared, evaluated, debated, and countered with alternative explanations. During argumentation, students build on and critique the explanations of their peers. Using the C-E-R framework allows them to first

## Table 6.6. Practice 6: Constructing Explanations and Designing Solutions

**Practice 6: Constructing Explanations and Designing Solutions**

| Grades K–2 | Grades 3–5 | Grades 6–8 | Grades 9–12 |
|---|---|---|---|
| Constructing explanations and designing solutions in K–2 builds on prior experiences and progresses to the use of evidence and ideas in constructing evidence-based accounts of natural phenomena and designing solutions.<br>• Make observations (firsthand or from media) to construct an evidence-based account for natural phenomena.<br>• Use tools and/or materials to design and/or build a device that solves a specific problem or a solution to a specific problem.<br>• Generate and/or compare multiple solutions to a problem. | Constructing explanations and designing solutions in 3–5 builds on K–2 experiences and progresses to the use of evidence in constructing explanations that specify variables that describe and predict phenomena and in designing multiple solutions to design problems.<br>• Construct an explanation of observed relationships (e.g., the distribution of plants in the backyard).<br>• Use evidence (e.g., measurements, observations, patterns) to construct or support an explanation or design a solution to a problem.<br>• Identify the evidence that supports particular points in an explanation.<br>• Apply scientific ideas to solve design problems.<br>• Generate and compare multiple solutions to a problem based on how well they meet the criteria and constraints of the design solution. | Constructing explanations and designing solutions in 6–8 builds on K–5 experiences and progresses to include constructing explanations and designing solutions supported by multiple sources of evidence consistent with scientific ideas, principles, and theories.<br>• Construct an explanation that includes qualitative or quantitative relationships between variables that predicts and/or describes phenomena.<br>• Construct an explanation using models or representations.<br>• Construct a scientific explanation based on valid and reliable evidence obtained from sources (including students' own experiments) and the assumption that theories and laws that describe the natural world operate today as they did in the past and will continue to do so in the future.<br>• Apply scientific ideas, principles, and/or evidence to construct, revise, and/or use an explanation for real-world phenomena, examples, or events.<br>• Apply scientific reasoning to show why the data or evidence is adequate for the explanation or conclusion.<br>• Apply scientific ideas or principles to design, construct, and/or test a design of an object, tool, process, or system.<br>• Undertake a design project, engaging in the design cycle, to construct and/or implement a solution that meets specific design criteria and constraints.<br>• Optimize performance of a design by prioritizing criteria, making tradeoffs, testing, revising, and re-testing. | Constructing explanations and designing solutions in 9–12 builds on K–8 experiences and progresses to explanations and designs that are supported by multiple and independent student-generated sources of evidence consistent with scientific ideas, principles, and theories.<br>• Make a quantitative and/or qualitative claim regarding the relationship between dependent and independent variables.<br>• Construct and revise an explanation based on valid and reliable evidence obtained from a variety of sources (including students' own investigations, models, theories, simulations, peer review) and the assumption that theories and laws that describe the natural world operate today as they did in the past and will continue to do so in the future.<br>• Apply scientific ideas, principles, and/or evidence to provide an explanation of phenomena and solve design problems, taking into account possible unanticipated effects.<br>• Apply scientific reasoning, theory, and/or models to link evidence to the claims to assess the extent to which the reasoning and data support the explanation or conclusion.<br>• Design, evaluate, and/or refine a solution to a complex real-world problem, based on scientific knowledge, student-generated sources of evidence, prioritized criteria, and tradeoff considerations. |

*Source:* Appendix F: Science and Engineering Practices in the *NGSS*. NGSS Lead States 2013, p. 61.

construct their explanations so that they can discuss, debate, evaluate, and modify them based on evidence, with the goal being to convince others of the validity of their claim and evidence in order to build a common understanding. Through argumentation, students often strengthen their explanations.

### Initiate, Respond, Evaluate (IRE) Method

Engaging in argument from evidence is the "minds-on" part of "hands-on, minds-on" duo we have been advocating so strongly. Many classroom discussions follow what has come to be known as the IRE method. The teacher initiates the question; the students respond and the teacher evaluates the responses. Even if students speak to each other, there is still the problem that students, instead of persuading, defer to the teacher, who "knows the answer" already. However, argumentation requires that students have a completely different goal in mind when they argue from evidence and try to convince their audience of the validity of their ideas.

*Scientific Argumentation in Biology: 30 Classroom Activities* (Sampson and Schleigh 2013) defines argumentation as a practice in science that is "an attempt to validate or refute a claim on the basis of reasons in a manner that reflects the values of the scientific community" (p. ix). In their book they provide 10 examples each of three types of instructional models for biology teachers; however, teachers of all subjects at all levels can use or modify these models to fit their teaching context. The instructional activities students engage in to build their practice of argumentation can be summarized as follows: (1) Generate Argument (students develop arguments based on supplied data); (2) Evaluate Alternatives (requires students to collect data appropriate for testing the efficacy of two or more alternatives); and (3) Write Refutational Essays (focus on discrediting naive ideas in favor of those accepted by the scientific community). This book also provides rubrics for providing feedback or evaluating the work of students.

Expecting students to argue for collaborative results without taking the time to teach them the norms and conventions of engaging in argumentation is asking for disappointment. Students who are arguing out loud with other students are sometimes performing for the teacher. Students need to be included, up front, in the purpose of their arguments and explanations and to see their fellow classmates as the audience with whom they want to reach consensus. They need to see that the purpose of argumentation is not winning but working together, using data and evidence to inform and persuade an audience that may see things differently than they do. Just as in science, evidence is the arbiter in changing minds and creating knowledge.

The *NGSS* and *Framework* explicitly identify argumentation as one of the eight scientific practices. However, even before argumentation recently came to the forefront as a practice of science, many of us have been engaging teachers and students in argumentation in professional development courses, classes at the university level, in schools, and in online courses and webinars. We both remember students making claims, and then saying, "Now that I hear myself saying that out loud, I can hear how weak my argument is." Our experience tells us that hearing oneself say things out loud is helpful because it brings the inner voice out into the open where it can be evaluated by self and others. We also realize that arguing in a group is a skill that needs to be learned. Students are used to arguing to win rather than to reach consensus. Taking the time to establish norms and conventions for productive talk cannot be emphasized enough.

## Author Vignette

In one teacher workshop Page and I did together, a participant asked if we had ideas on how to help students argue without malice and do so constructively. We responded by suggesting that "talking in class" be taught since in many classrooms in the past, students only spoke when spoken to. We suggested that the teacher and students set up a "bill of rights" for speakers and rules for listening politely and not interrupting. We have also found success in asking students to respond to one another by repeating what the first student has said so the former knows that he or she was heard properly. Then the responder could enter his or her own idea or critique of the former student's claim or argument. Another technique we shared is to have the responder say something positive about the speaker's comments and then express a lack of understanding with the part of the speaker's claim that is confusing or with which the responder disagrees.

—Dick Konicek-Moran

We stress respect and use the techniques offered by Karen Gallas (1995) or through TERC's Talk Science project (*http://inquiryproject.terc.edu*). Gallas describes a class that did not go well because certain students dominated the conversations. She spoke with them and described the problem, and they responded by backing off and helping other students to "get into the dialogue." We realize that it takes time for students to get used to the "don't need to raise your hand" model of conversational argumentation, and we realize as many students are reluctant to participate as are anxious to dominate. One of the most difficult parts of free argumentation in a classroom is for the teacher to vacate the role of having to validate each response and to just become part of the conversation. However, students are helped by teachers who encourage questions such as: "How do you know?" or "What evidence do you have to back up your claim?" Teachers need to model such behavior.

Most importantly, students build conceptual understanding through argumentation. As they frame their arguments, they solidify their own conceptual understanding. In the process of evaluating the arguments of others, they often wrestle with their own ideas and reconstruct their thinking based on others presenting new evidence or reasoning that makes more sense than one's own justification.

The media often are guilty of making nonscientific claims and statements. Recently the public was told that the Moon was going to be two to three times its original size on a particular date. Sometimes unscientific polls are taken and broadcast to the nation, asking about opinions or expectations. These are too often taken as fact and in some cases can affect voting habits or the outcome of elections. Some political parties will tell us that their candidate is much closer in polls than is true in order to get voters to turn out. We have debates among political opponents that are filled with "facts" that, after being checked, show great discrepancies between them and the truth. Unfortunately, we do not usually have an opportunity to engage in argumentation with them ourselves. Individuals who use such superficial criteria for decision making choose our political leaders in democracies all over the world. Scientifically literate people need to be able to carefully evaluate the claims of others and press for an evidence-based argument and not just listen to and accept a set of talking points.

Perhaps science education is a way of helping solve these problems. Political affiliations and familial connections will always be strong but we must try to help our next generation to listen, ask good questions, and demand evidence for the "facts" that are presented to them. We need to develop an atmosphere of alert skepticism, which stops short of cynicism.

*NGSS* Appendix F (Science and Engineering Practices in the *NGSS*) includes a matrix of the progression of what students are expected to do for the practice of Engaging in Argument From Evidence at different grade spans. As you examine Table 6.7, notice how this practice overlaps with other practices.

### *Obtaining, Evaluating, and Communicating Information*

*A Framework for K–12 Science Education* summarizes this science practice as follows:

> *Science cannot advance if scientists are unable to communicate their findings clearly and persuasively or to learn about the findings of others. A major practice of science is thus the communication of ideas and the results of inquiry—orally, in writing, with the use of tables, diagrams, graphs, and equations, and by engaging in extended discussions with scientific peers. Science requires the ability to derive meaning from scientific texts (such as papers, the internet, symposia, and lectures), to evaluate the scientific validity of the information thus acquired, and to integrate that information.* (NRC 2012, p. 53)

Whether your students are heading toward careers in science, engineering, or just plain citizenry, they need to be able to be literate in many fields, including science. Scientific literacy includes not only the knowledge and understanding of science concepts and the way science is practiced; it also includes reading and writing to understand science concepts as well as writing and speaking to communicate to a particular audience. Scientists read and write extensively to help themselves better understand the scientific ideas they are researching and communicate them to a specific audience. Likewise, students should

## Table 6.7. Practice 7: Engaging in Argument From Evidence

**Practice 7: Engaging in Argument from Evidence**

| Grades K–2 | Grades 3–5 | Grades 6–8 | Grades 9–12 |
|---|---|---|---|
| Engaging in argument from evidence in K–2 builds on prior experiences and progresses to comparing ideas and representations about the natural and designed world(s).<br>• Identify arguments that are supported by evidence.<br>• Distinguish between explanations that account for all gathered evidence and those that do not.<br>• Analyze why some evidence is relevant to a scientific question and some is not.<br>• Distinguish between opinions and evidence in one's own explanations.<br>• Listen actively to arguments to indicate agreement or disagreement based on evidence, and/or to retell the main points of the argument.<br>• Construct an argument with evidence to support a claim.<br>• Make a claim about the effectiveness of an object, tool, or solution that is supported by relevant evidence. | Engaging in argument from evidence in 3–5 builds on K–2 experiences and progresses to critiquing the scientific explanations or solutions proposed by peers by citing relevant evidence about the natural and designed world(s).<br>• Compare and refine arguments based on an evaluation of the evidence presented.<br>• Distinguish among facts, reasoned judgment based on research findings, and speculation in an explanation.<br>• Respectfully provide and receive critiques from peers about a proposed procedure, explanation, or model by citing relevant evidence and posing specific questions.<br>• Construct and/or support an argument with evidence, data, and/or a model.<br>• Use data to evaluate claims about cause and effect.<br>• Make a claim about the merit of a solution to a problem by citing relevant evidence about how it meets the criteria and constraints of the problem. | Engaging in argument from evidence in 6–8 builds on K–5 experiences and progresses to constructing a convincing argument that supports or refutes claims for either explanations or solutions about the natural and designed world(s).<br>• Compare and critique two arguments on the same topic and analyze whether they emphasize similar or different evidence and/or interpretations of facts.<br>• Respectfully provide and receive critiques about one's explanations, procedures, models, and questions by citing relevant evidence and posing and responding to questions that elicit pertinent elaboration and detail.<br>• Construct, use, and/or present an oral and written argument supported by empirical evidence and scientific reasoning to support or refute an explanation or a model for a phenomenon or a solution to a problem.<br>• Make an oral or written argument that supports or refutes the advertised performance of a device, process, or system based on empirical evidence concerning whether or not the technology meets relevant criteria and constraints.<br>• Evaluate competing design solutions based on jointly developed and agreed-upon design criteria. | Engaging in argument from evidence in 9–12 builds on K–8 experiences and progresses to using appropriate and sufficient evidence and scientific reasoning to defend and critique claims and explanations about the natural and designed world(s). Arguments may also come from current scientific or historical episodes in science.<br>• Compare and evaluate competing arguments or design solutions in light of currently accepted explanations, new evidence, limitations (e.g., tradeoffs), constraints, and ethical issues.<br>• Evaluate the claims, evidence, and/or reasoning behind currently accepted explanations or solutions to determine the merits of arguments.<br>• Respectfully provide and/or receive critiques on scientific arguments by probing reasoning and evidence, challenging ideas and conclusions, responding thoughtfully to diverse perspectives, and determining additional information required to resolve contradictions.<br>• Construct, use, and/or present an oral and written argument or counter-arguments based on data and evidence.<br>• Make and defend a claim based on evidence about the natural world or the effectiveness of a design solution that reflects scientific knowledge and student-generated evidence.<br>• Evaluate competing design solutions to a real-world problem based on scientific ideas and principles, empirical evidence, and/or logical arguments regarding relevant factors (e.g., economic, societal, environmental, ethical considerations). |

*Source:* Appendix F: Science and Engineering Practices in the *NGSS*. NGSS Lead States 2013, p. 63.

also have many opportunities to read informational text, write about scientific ideas and findings, and communicate their scientific ideas in a variety of ways to a specific audience. Students also need to learn how to be critical consumers of information, whether they read it in a report or book, hear it on the news, or find information on the internet.

*A Framework for K–12 Science Education* describes the goals for helping students meet the practice of Obtaining, Evaluating, and Communicating Information. By the end of grade 12, all students should be able to do the following:

- Use words, tables, diagrams, and graphs (whether in hard copy or electronically), as well as mathematical expressions, to communicate their understanding or to ask questions about a system under study.

- Read scientific and engineering text, including tables, diagrams, and graphs, commensurate with their scientific knowledge and explain the key ideas being communicated.
- Recognize the major features of scientific and engineering writing and speaking and be able to produce written and illustrated text or oral presentations that communicate their own ideas and accomplishments.
- Engage in a critical reading of primary scientific literature (adapted for classroom use) or of media reports of science and discuss the validity and reliability of the data, hypotheses, and conclusions (NRC 2012, pp. 74–75).

A recent article in *Science Scope* (Bell et al. 2012) describes an example of what this practice looks like in a grade 10 biology classroom in which students participated in contemporary infectious disease research: Students learned the biology behind why various pathogens make humans sick at the cellular level, as well as the science behind how and why infectious diseases are transmitted locally and globally. Students had their choice of project: a local social network analysis to learn about and apply constructs like herd immunity or a global epidemic modeling study to explore the various factors affecting the spread of infectious disease, such as seasonality and viral latency periods. As part of these projects, students read original research, communicated with scientists who conduct this type of research, and conducted their own research. Students developed products to communicate various aspects of their work to scientists and other health professionals, their teachers, and their peers. These products included a research design plan, an elevator speech, and an original research paper.

A recent article in *Science and Children,* "Negotiating the Way to Inquiry" (Kuhn and McDermott 2013), emphasizes the practice of communication. Much of the methodology in their article was based on the model of teaching called the "Science Writing Heuristic" developed by Norton-Meier and colleagues (2008) that focused on incorporating creativity into communication about science concepts and stressing negotiation. In this article, the authors point out that students have a great many opportunities to write lab reports or fill in worksheets (for teacher evaluation) but seldom find a way to communicate with different audiences, particularly in a multimodal way. They tell about fourth graders writing for second graders so their audience could understand their project. This provided the opportunity to communicate their science knowledge in an appropriate level and, at the same time, helped them understand their own meaning making. We all know that we learn so much more when preparing to teach.

A well-conceived study by two Iowa professors, Michael McDermott and Brian Hand, concentrated on a reanalysis of studies on science writing over a 10-year period (2010). It was obvious that having students write for an audience other than their teacher was perceived positively by a vast majority of the students in the studies. Their comments consisted of (paraphrased) statements such as:

> *Writing for someone else than my teacher made me use language other than the language my teacher used. I had to use language that would make sense to my audience and in doing so I learned a lot about my own understanding.*

> and

> *Thinking over and over again about what I was trying to teach someone made me realize that I needed to understand things better, myself. I also realized that I needed to put in some drawings too.*

We have found that asking children to write specifically about what they did not understand was very helpful and often as they wrote, they ended up answering their own questions and clearing up their confusions.

So often, writing in science involves mainly writing formal lab reports using a specified format. But, there are so many other ways students can write in science. Whatever writing task you use, the students need to be part of the innovation. Not only will their metacognitive skills and epistemological understanding become stronger, but they will also become partners in the scientific community you are trying to build.

*NGSS* Appendix F (Science and Engineering Practices in the *NGSS*) includes a matrix of the progression of what students are expected to do for the practice of Obtaining, Evaluating, and Communicating Information at different grade spans. As you examine Table 6.8 (p. 136), notice how this practice overlaps with other practices.

*A Framework for K–12 Science Education* includes this reflection on the practices:

> *Our view is that the opportunity for students to learn the basic set of practices is also an opportunity to have them stand back and reflect on how these practices contribute to the accumulation of scientific knowledge. For example, students need to see that the construction of models is a major means of acquiring new understanding; that these models identify key features and are akin to a map, rather than a literal interpretation of reality; and that the great achievement of science is a core set of explanatory theories that have wide application. (NRC 2013, p. 78)*

Our reflection on these practices is that they support conceptual learning and understanding and that neither should be taught and learned in isolation of the other.

## Table 6.8. Practice 8: Obtaining, Evaluating, and Communicating Information

**Practice 8: Obtaining, Evaluating, and Communicating Information**

| Grades K–2 | Grades 3–5 | Grades 6–8 | Grades 9–12 |
|---|---|---|---|
| Obtaining, evaluating, and communicating information in K–2 builds on prior experiences and uses observations and texts to communicate new information.<br>• Read grade-appropriate texts and/or use media to obtain scientific and/or technical information to determine patterns in and/or evidence about the natural and designed world(s).<br>• Describe how specific images (e.g., a diagram showing how a machine works) support a scientific or engineering idea.<br>• Obtain information using various texts, text features (e.g., headings, tables of contents, glossaries, electronic menus, icons), and other media that will be useful in answering a scientific question and/or supporting a scientific claim.<br>• Communicate information or design ideas and/or solutions with others in oral and/ or written forms using models, drawings, writing, or numbers that provide detail about scientific ideas, practices, and/or design ideas. | Obtaining, evaluating, and communicating information in 3–5 builds on K–2 experiences and progresses to evaluating the merit and accuracy of ideas and methods.<br>• Read and comprehend grade-appropriate complex texts and/or other reliable media to summarize and obtain scientific and technical ideas and describe how they are supported by evidence.<br>• Compare and/or combine across complex texts and/or other reliable media to support the engagement in other scientific and/or engineering practices.<br>• Combine information in written text with that contained in corresponding tables, diagrams, and/or charts to support the engagement in other scientific and/or engineering practices.<br>• Obtain and combine information from books and/or other reliable media to explain phenomena or solutions to a design problem.<br>• Communicate scientific and/or technical information orally and/or in written formats, including various forms of media as well as tables, diagrams, and charts. | Obtaining, evaluating, and communicating information in 6–8 builds on K–5 experiences and progresses to evaluating the merit and validity of ideas and methods.<br>• Critically read scientific texts adapted for classroom use to determine the central ideas and/or obtain scientific and/or technical information to describe patterns in and/or evidence about the natural and designed world(s).<br>• Integrate qualitative and/or quantitative scientific and/or technical information in written text with that contained in media and visual displays to clarify claims and findings.<br>• Gather, read, and synthesize information from multiple appropriate sources and assess the credibility, accuracy, and possible bias of each publication and methods used, and describe how they are supported or not supported by evidence.<br>• Evaluate data, hypotheses, and/or conclusions in scientific and technical texts in light of competing information or accounts.<br>• Communicate scientific and/or technical information (e.g., about a proposed object, tool, process, system) in writing and/or through oral presentations. | Obtaining, evaluating, and communicating information in 9–12 builds on K–8 experiences and progresses to evaluating the validity and reliability of the claims, methods, and designs.<br>• Critically read scientific literature adapted for classroom use to determine the central ideas or conclusions and/or to obtain scientific and/or technical information to summarize complex evidence, concepts, processes, or information presented in a text by paraphrasing them in simpler but still accurate terms.<br>• Compare, integrate, and evaluate sources of information presented in different media or formats (e.g., visually, quantitatively) as well as in words in order to address a scientific question or solve a problem.<br>• Gather, read, and evaluate scientific and/ or technical information from multiple authoritative sources, assessing the evidence and usefulness of each source.<br>• Evaluate the validity and reliability of, and/ or synthesize, multiple claims, methods, and/or designs that appear in scientific and technical texts or media reports, verifying the data when possible.<br>• Communicate scientific and/or technical information or ideas (e.g., about phenomena and/or the process of development and the design and performance of a proposed process or system) in multiple formats (i.e., orally, graphically, textually, mathematically). |

*Source:* Appendix F: Science and Engineering Practices in the *NGSS*. NGSS Lead States 2013, p. 65.

### Learning Strands

In addition to the three dimensions (disciplinary core ideas, crosscutting concepts, and scientific and engineering practices) recent research and advances in considering how students learn have outlined four *learning strands.* The learning strands are

- understanding science explanations;
- generating scientific evidence;
- reflecting on scientific knowledge; and
- participating productively in science.

In *Taking Science to School*, the strands are described as follows:

> *The strands are not independent or separable in the practice of science, nor in the teaching and learning of science. Rather, the strands of scientific proficiency are interwoven and*

> *taken together, are viewed as science as practice. … The science-as-practice perspective invokes the notion that learning science involves learning a system of interconnected ways of thinking in a social context to accomplish the goal of how conceptual understanding of natural systems is linked to the ability to develop explanations of phenomena and to carry out empirical investigations in order to develop or evaluate knowledge claims. (NRC 2007, pp. 36–38)*

The strands, which are relatively self-explanatory, are not to be construed as a formula for teaching. Research suggests that strong attention be paid to the strands as one develops lessons and curricula. They should be "woven" together so that they become common "practice" in the classroom. In the past, teachers would often teach "process skills" at the beginning of the year, separate from the content students would be learning, and then use the skills throughout the year as they teach content. Today we know that students learn science best when content and practices are taught together so students can see how science generates new knowledge and ways of thinking about our world. (See the appendix on p. 221 for a case study that demonstrates how all of these principles look in an actual classroom.)

## Questions for Reflection and Discussion

1. Try your hand at writing some refutational text for a topic of your choice. Try it out with your class or with a group of people and ask them about their opinion.
2. View the video titled "Completing the Circuit," from PBS Learning Media (*www.pbslearningmedia.org/resource/tdpd12.pd.sci.compcirc/completing-the-circuit*). Analyze the teaching against the practices mentioned in the *NGSS*.
3. Read the article "Negotiating the Way to Inquiry" in *Science and Children* (Kuhn and McDermott 2013). Do you think this article presents a reasonable way to have students understand the topic about which they are writing?
4. Do you believe that teaching science-as-practice will result in better science teaching than past attempts to include science as a process? Why or why not?
5. Which of the practices do you feel you need to more fully develop in your teaching? How do you plan to do this?
6. How do you or would you teach civility in argumentation in your classroom? What types of norms would you develop with your students?
7. In the article "Do Plants Drink Green Water?" in *Educational Leadership* (Beck and Leishman 1996), what do you think the teacher means by being amazed about how much power she has in the classroom?
8. Choose one "golden line" from this chapter (a sentence that really speaks to or resonates with you). Write this on a sentence strip and share it with others. Explain why you chose it.

9. What was the biggest "takeaway" from this chapter for you? What will you do or think about differently as a result?

## Extending Your Learning With NSTA Resources

1. NSTA has organized all of its *NGSS* resources at *http://ngss.nsta.org*. At this site, you can find books, journal articles, videos, web seminars, and other resources to extend your learning about *NGSS* and the scientific and engineering practices.
2. Read and discuss Chapter 2: The Four Strands of Learning, in Michaels, S., A. Shouse, and H. Schweingruber. 2008. *Ready, set, SCIENCE! Putting research to work in K–8 science classrooms.* Washington, DC: National Academies Press.

# Chapter 7

## How Does the Use of Instructional Models Support Teaching for Conceptual Understanding?

As our brief history of science education in Chapter 5 implied, instructional models have long been a feature of curriculum materials. What is the difference between an instructional model, instructional strategies, and a curriculum, you might ask? A curriculum is more or less a roadmap of what is to be taught and learned, instructional strategies are pedagogical techniques, and an instructional model is a framework of how the instruction will be presented. "Although well-designed and sequential instructional materials can enhance student learning, they cannot do it all. The interaction between teachers and students is the key. Incorporating an instructional model into the instructional materials brings the materials as close as possible to facilitating the best interaction between teachers and students" (Bybee 2010, p. 11).

Rogoff and colleagues suggest three types of instructional models: the adult-led model, the child-led (student-led) model, and the community of learners model (Rogoff, Matusov, and White 1996). They argue that the first two models are either/or models and that a community of learners model provides both teacher and student the opportunity to contribute to the acquisition of knowledge.

Many instructional models have been promoted during the last 50 years. Some are still standing while others have either been modified or gone by the wayside. The three best-known models we will focus on are the Learning Cycle Model, the 5E Instructional Model, and the Conceptual Change Model (CCM). Both of us have been actively teaching students and teachers for the five decades that have produced these models, so we've had a chance to use them, evaluate them, and modify them.

### The Learning Cycle Instructional Model

Probably no other model has had so many iterations and variations over the past five decades than the *learning cycle*. They have some commonalities: first, they assume student activity; second, they expect some input by the teacher or guide regarding science concepts; and lastly, they include a focus on application of the concept learned in a different context.

The first time the term learning cycle was alluded to in print was in 1962 when J Myron Atkin, from Stanford University, and the late Robert Karplus, then from the Berkeley School of Education wrote an article for *The Science Teacher* entitled, "Discovery or Invention?" (1962). As the title suggests, the article asked if students should be asked to discover the "truths" of science or to personally invent them through guided inquiry. Their theory was based strongly upon Piaget and his stage theory of development (1973). The curriculum that eventually grew from this model was the Science Curriculum Instructional Strategy (SCIS), in which this particular cycle was polished (1970–1974). Bob Karplus published an article in 1977 in which he laid out the parts of the cycle in detail:

1. Exploration: students explore on their own with minimal guidance
2. Concept introduction (invention): the concept is introduced and explained by the teacher
3. Concept application: the concept is applied to a new situation and its range of applicability is extended

In practice, it might have looked something like this: Each student in a class is given a lightbulb, a wire, and a battery and asked to find as many ways as they can to light the bulb (Figure 7.1). This poses a problem for many people since their only experience with these items may have been in a flashlight or in a physics lab with sockets and switches. If you find this hard to believe, see the Private Universe Project video *Minds of Our Own, Can We Believe Our Eyes?* to see how Harvard and Massachusetts Institute of Technology (MIT) graduates fail to light a bulb using a wire and a battery (1997). This struggle is the *exploration* phase of the cycle. Once someone is successful, the methods travel like wildfire around the classroom. Soon, the classroom looks like a birthday cake alive with brightly lit bulbs.

**Figure 7.1. Four Ways to Light a Bulb**

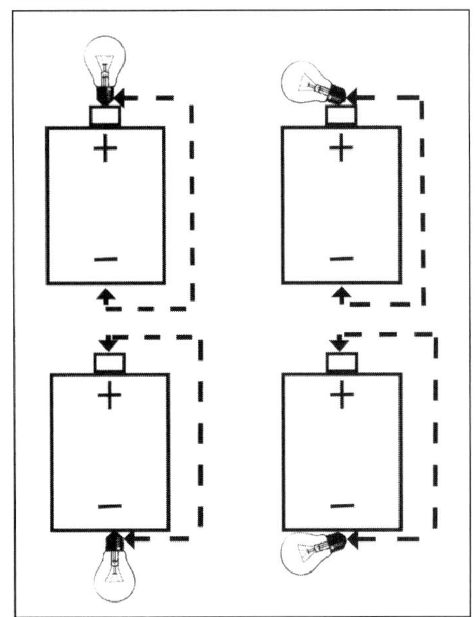


After the class has mastered the task with guidance from the teacher and their peers, the class and teacher "invent" the concept of a *complete circuit*. This is the second phase of the cycle, the *concept invention* (or concept development).

The students are next asked to try various types of circuits, explore parallel and series circuits, and apply the concept of *circuit* to various new situations. They may be asked to wire a dollhouse with switches in each room or floor using batteries as power or to build a car that operates on battery power and tiny motors. Children can make switches from foil wrap or build flashlights or bulb sockets from easily obtainable materials.

Critics found that this early model assumed that children's naive ideas didn't matter; that teachers were able to be good diagnosticians of student's ideas and that the "discovery" of the concept would just happen (NRC 2007).

John Renner of the University of Oklahoma opted for a learning cycle model that relied on relating the concept information to the students, confirming and then practicing the information through activity, and then applying the concepts through solving or elaborating on problems devised by the teacher or a textbook. This model also overlooked the importance of students' prior knowledge, and suggested the students not be allowed to stray from the prescribed path of instruction—in other words, that they not "mess about."

Many other versions of the original Karplus learning cycle model have been developed. A common feature of each of these modified versions is a focus on conceptual reconstruction, following a growing recognition of the importance of students' preconceptions. The following chart (Table 7.1) summarizes the variety of learning cycle frameworks proposed by different researchers for changing students' ideas adapted from Karplus's original three phases: exploration, concept invention, and concept application (Sunal 1992). Note the similarities as well as differences between them. Many instructional materials today use variations of these models.

## Table 7.1. Summary of the Learning Cycle Frameworks

| Model | Phase 1 | Phase 2 | Phase 3 | Phase 4 |
|---|---|---|---|---|
| Renner Model | Experiences | Interpretation | Exploration | |
| Driver Model | Discovery | Presentation | Application | |
| Nussbaum and Novick Model | Expose alternative frameworks | Create conceptual conflict | Encourage cognitive accommodation | |
| Barnes Model | Focusing | Exploration | Reorganization | Public sharing |
| Rowell and Dawson Model | Establish initial ideas | Introduce new ideas | Comparison of ideas | |
| Riverina and Murray Model | Identify naive ideas. Select events. | Exploratory activities | Organize ideas and establish links | Practice and apply new idea |
| Hewson and Hewson Model | Diagnose | Opportunity to clarify and contrast | Practice new idea | Apply new idea |
| Lawson and Abraham Model | Exploration | Conceptual invention | Expansion | |

## Author Vignette

In the late 1990s I had the opportunity to join a state team of Maine physics and physical science teachers to learn about and pilot instructional materials being developed with National Science Foundation (NSF) funding by Dr. Fred Goldberg's group at San Diego State University, *Constructing Physics Understanding* (CPU). In San Diego we learned about the CPU pedagogy, which was similar to the learning cycle model, and we followed the CPU instructional model to work through and test out the instructional materials. The CPU instructional model consists of three phases: elicitation, development, and application. During the elicitation phase, we were presented with an interesting phenomenon in which we made predictions about the outcome and actively engaged in small-group and whole-class discussions about the outcome before and after making observations. What made this model unique at that time, compared to all the previous learning cycle models, was the use of computer-based simulations for the development phase. As we worked through the computer simulations, we recast several of our initial ideas, and special software kept track of our evolving ideas. Through whole-group discussions we constructed and reconstructed our own understanding of the concept. I was struck with how powerful the use of technology tools could be when incorporated into a learning cycle model and combined with rich discourse. The application phase provided examples of ways in which students could practice and apply the ideas they formally developed through evidence-based discussions. My experience as a learner with the CPU project reinforced for me the importance of using an instructional model that parallels the phases we go through as learners during the learning process.

—Page Keeley

## 5E Instructional Model

In 1989, Rodger Bybee, then the director of the Biological Sciences Curriculum Study (BSCS), developed arguably the most ubiquitous version of the learning cycle, the 5E Instructional Model (Bybee et al. 1989; see Table 7.2).

### Table 7.2. Summary of the BSCS 5E Instructional Model

| Phase | Summary |
|---|---|
| Engagement | The teacher or a curriculum task assesses the learners' prior knowledge and helps them become engaged in a new concept through the use of short activities that promote curiosity and elicit prior knowledge. The activity should make connections between past and present learning experiences, expose prior conceptions, and organize students' thinking toward the learning outcomes of current activities. |
| Exploration | Exploration experiences provide students with a common base of activities within which current concepts (i.e., misconceptions), processes, and skills are identified and conceptual change is facilitated. Learners may complete lab activities that help them use prior knowledge to generate new ideas, explore questions and possibilities, and design and conduct a preliminary investigation. |
| Explanation | The explanation phase focuses students' attention on a particular aspect of their engagement and exploration experiences and provides opportunities to demonstrate their conceptual understanding, process skills, or behaviors. This phase also provides opportunities for teachers to directly introduce a concept, process, or skill. Learners explain their understanding of the concept. An explanation from the teacher or the curriculum may guide them toward a deeper understanding, which is a critical part of this phase. |
| Elaboration | Teachers challenge and extend students' conceptual understanding and skills. Through new experiences, the students develop deeper and broader understanding, more information, and adequate skills. Students apply their understanding of the concept by conducting additional activities. |
| Evaluation | The evaluation phase encourages students to assess their understanding and abilities and provides opportunities for teachers to evaluate student progress toward achieving the educational objectives. |

*Source:* Bybee 2009, p. 5.

As you can see from the table above, there is a clear intent to help teachers plan lessons that include inquiry and the scientific practices described by *A Framework for K–12 Science Education* and included in the *NGSS*. The teacher is still expected to be an instructional architect and must know the subject matter well. The 5E Model remains the most popular model in science teaching and is found in most, if not all, of the 21st-century treatises on teaching science.

Furthermore, the 5E Model is supported by key findings from the seminal publication on the research on how people learn science, *How People Learn* (Bransford, Brown, and Cocking 2000). The authors say this of the use of the 5E Model:

> *An alternative to simply progressing through a series of exercises that derive from a scope and sequence chart is to expose students to the major features of a subject domain as they arise naturally in problem situations. Activities can be structured so that students are*

## Figure 7.2. The 7E Model

FIGURE 2

**Seatbelt lesson using the 7E model.**

**Elicit prior understandings**

- Students are asked, "Suppose you had to design seat belts for a racecar traveling at high speeds. How would they be different from ones available on passenger cars?" The students are required to write a brief response to this "What do you think?" question in their logs and then share with the person sitting next to them. The class then listens to some of the responses. This requires a few minutes of class time.

**Engage**

- Students relate car accidents they have witnessed in movies or in real life.

**Explore**

- The first part of the exploration requires students to construct a clay figure they can sit on a cart. The cart is then crashed into a wall. The clay figure hits the wall.

**Explain**

- Students are given a name for their observations. Newton's first law states, "Objects at rest stay at rest; objects in motion stay in motion unless acted upon by a force."

**Engage**

- Students view videos of crash test dummies during automobile crashes.

**Explore**

- Students are asked how they could save the clay figure from injury during the crash into the wall. The suggestion that the clay figure will require a seat belt leads to another experiment. A thin wire is used as a seat belt. The students construct a seat belt from the wire and ram the cart and figure into the wall again. The wire seat belt keeps the clay figure from hitting the wall, but the wire slices halfway through the midsection.

**Explain**

- Students recognize that a wider seatbelt is needed. The relationship of pressure, force, and area is introduced.

**Elaborate**

- Students then construct better seat belts and explain their value in terms of Newton's first law and forces.

**Evaluate**

- Students are asked to design a seat belt for a racing car that travels at 250 km/h. They compare their designs with actual safety belts used by NASCAR.

**Extend**

- Students are challenged to explore how airbags work and to compare and contrast airbags with seat belts. One of the questions explored is, "How does the airbag get triggered? Why does the airbag not inflate during a small fender-bender but does inflate when the car hits a tree?"

*Source:* Eisenkraft 2003.

*able to explore, explain, extend, and evaluate their progress. Ideas are best introduced when students see a need or a reason for their use—this helps them see relevant uses of knowledge to make sense of what they are learning. (Bransford, Brown, and Cocking 2000, p. 127)*

Additionally, the 5E Model has been shown to be effective in teacher professional development (Bybee 2010).

Recently the 5E Model was modified to include the elicitation of prior student ideas within the *Engagement* stage, presumably eliminating the need for certain *E*s to be added to the model. However, merely eliciting the students' prior ideas is not enough. There should be some follow up on those ideas with discussion among the members of the classroom community. This will help the teacher analyze the prior conceptions and plan for the engagement phase. Otherwise, the teacher is following a preset plan without the formative assessment information about students' thinking driving instructional decisions. Some educators believe the *eliciting* of student prior ideas deserves an *E* of its own.

Arthur Eisenkraft, author and project director of *Active Physics* and past-president of NSTA, suggests expanding the 5E Model to a 7E Model (2003). He explains in his article in *The Science Teacher* that because instructors tend to omit crucial parts of the cycle, making them explicit may be helpful. Therefore, he suggests enlarging the *Engage* section to include both *Elicit* and *Engage* and to expand *Elaborate* and *Evaluate* into three parts: *Elaborate, Evaluate,* and *Extend* (see Figure 7.2). Eisenkraft points out that one of his main goals in expanding the model is to promote transfer of learning particularly in the *Elaboration* phase. He mentions explicitly the progression from *near transfer of learning* to *distant transfer of learning,* in which the context of transfer is not merely the redoing of an investigation with a similar substance and calculations, but a jump to a distant concept such as using the concept of *phase change* (the change from one state to another, in this case the change from a stationary object to a moving object) to explain traffic congestion. Now that would be a far reach for many people! Although, he does not mention it specifically, the area of elaboration to a distant transfer also fits beautifully into an introduction to an engineering problem. Figure 7.2 shows an example of using the 7E Model for a force and motion lesson.

Dr. Emilio Duran and his colleagues describe another modification of the 5E Model in his article "A Learning Cycle for All Students" (Duran, Duran, Haney, and Scheuermann 2011). They suggest inserting a conscious pause, called the *Express* stage, between *Explain* and *Elaborate* to formatively assess and ensure that *all* students are progressing adequately through the early stages of the 5E Model. Formative assessment is used at this stage to inform and guide whether some students may need additional differentiated instruction during the *Elaborate* phase by providing an opportunity for students to safely express their ideas at this point, which may uncover conceptual misunderstandings that might not be noted by the teacher until later in the cycle. In this way, appropriate instructional experiences can be selected for targeted students during the *Elaborate* stage.

## The Conceptual Change Model

In the previous chapters we described concepts, conceptual change, and preconceptions—and the pitfalls and opportunities provided by these preconceptions. The Conceptual Change Model (CCM) is based on constructivist theory, i.e., that knowledge cannot be directly transmitted from one person to another and that knowledge is best achieved by personal construction, often ameliorated by social interaction. After a decade or more of educational research on children's alternative ideas or preconceptions, it behooved the educational community to make use of this information and concentrate on how these many alternative conceptions could be modified to fit the current beliefs of the scientific community. Several of the three- and four-stage learning cycle models in Table 7.1 (p. 141) can be considered conceptual change models. In 1982, a model for attaining conceptual change was developed at Cornell University by Posner, Strike, Hewson, and Gertzog. In their article "Accommodation of a Scientific Conception: Toward a Theory of Conceptual Change," they outlined four requirements necessary for a person to modify personal concepts (1982). They posited that in order for a person to modify their concepts

1. the learner must become dissatisfied with the concept,
2. the new concept must appear plausible,
3. the new concept must be reasonable and understandable, and
4. the new concept must have value and possess predictive power.

The first requirement makes sense, for why would a person want to change a conceptual model that is working perfectly well? The second and third requirements emphasize the point that the newly introduced concept must appear within the realm of believability to the student. And finally, the student would have to find the new concept more useful than the previous one, which harkens back to the first principle of satisfaction.

Their model was criticized for lack of their consideration of affect, particularly motivation. It appears that motivation is quite important, as any classroom teacher knows. If the teacher can achieve an environment of trust and provide challenges for students that affect their lives and can develop their curiosity about a task, the intentional conceptual change will follow (Sinatra and Pintrich 2003).

In December of 1982, Nussbaum and Novick published an article in *Instructional Science* that suggested a more instructionally based model than the one posed by Posner and colleagues. They suggested that the following sequence would help teachers to achieve conceptual change: First, the teacher's job is to reveal the students' preconceptions by bringing them into the open and then promote discussion so the preconceptions can be evaluated. They then propose that the teacher—by means of probing, questioning, or performing demonstrations such as discrepant events—point out some conflict(s) that arise if the preconceptions are to be considered useful. Then the teacher uses analogy, or any number of teaching strategies available (see the strategies emphasized in Chapter 8) to lead the students toward the new concept more in tune with current scientific beliefs (Nussbaum and Novick 1982). We particularly like the fact that this model advocates probing for student preconceptions and therefore uses formative assessment and promotes discussions and argumentation in tune with the *Next Generation Science Standards* (*NGSS*). The Conceptual Change Model (CCM) has undergone many revisions, with contributions from teachers, researchers, and curriculum developers. Joseph Stepans and colleagues describe a six-stage conceptual change model that expands on the original three-stage Atkin and Karplus learning cycle model and includes elements from Posner and colleagues' model for teaching for conceptual change. The six steps are as follows:

1. Commit to an outcome
2. Expose beliefs
3. Confront beliefs
4. Accommodate the concept
5. Extend the concept
6. Go beyond (Stepans, Saigo, and Ebert 1999, p. 141)

Particular features are present during conceptual change teaching that uses an instructional model such as the one above (Hewson 1992). These features include the following:

1. Metacognition: Students are encouraged or are able to step back from one or more ideas held by themselves or others in order to think about them and express their ideas about them.
2. Classroom climate: There is an attitude of respect by both teacher and students for the ideas of others, even when they are contradictory.
3. Role of the teacher: The teacher is able to provide opportunities for students to express themselves without fear of ridicule, and to ensure that he or she is not the sole determiner of what counts as an acceptable idea, explanation, or argument.
4. Role of the learner: Students are willing to take responsibility for their own learning, listen to and evaluate others ideas, and change their ideas when another seems more viable to them. Students monitor their own learning.

Hewson states that teaching that has included these components has been successful in helping elementary, middle school, high school, and college students significantly change their conceptions in many different areas in physics, chemistry, biology, and Earth science.

## A New Instructional Model: Argument-Driven Inquiry (ADI)

With the release of *A Framework for K–12 Science Education* and the *Next Generation Science Standards,* instruction is shifting from the pressure of "covering content" to providing increased opportunities for students to authentically participate in scientific practices that foster deeper learning of content. Laboratory experiences, defined as "opportunities for students to interact directly with the material world using the tools, data collection techniques, models, and theories of science" are an integral part of science learning (NRC 2005, p. 3). Traditional laboratory experiences follow the presentation of the content to be learned as verification activities, focus on procedural learning, and, according to research, actually do little to promote conceptual understanding of science that leads to the development of the knowledge and abilities needed to be proficient in science. Laboratory activities have been such a mainstay in science classrooms, particularly in high school science, that it is hard for teachers to accept one of the conclusions from *America's Lab Report* that "the quality of current laboratory experiences is poor for most students" (NRC 2005, p. 6). The report concluded that

> *Four principles of instructional design can help laboratory experiences achieve their intended learning goals if: (1) they are designed with clear learning outcomes in mind, (2) they are thoughtfully sequenced into the flow of classroom science instruction, (3) they are designed to integrate learning of science content with learning about the processes of science, and (4) they incorporate ongoing student reflection and discussion. (NRC 2005, p. 6)*

Dr. Victor Sampson and his colleagues have addressed the importance of ensuring that laboratory activities provide students with the opportunity to engage in the practices of science and develop deeper understanding of science through their 8-stage instructional model, Argument-Driven Inquiry (ADI), illustrated in Figure 7.3.

## Figure 7.3. Argument-Driven Inquiry

**Stage 1:** Identify the task and the guiding question. Hold a "tool talk"

Small groups of students then …

**Stage 2:** Design a method and collect data

Groups then …

**Stage 3:** Analyze data and develop a tentative argument

Each group then shares its argument during an …

**Stage 4:** Argumentation session

If needed, groups can …

Collect additional data or reanalyze the collected data

The teacher then leads an …

The teacher then leads an …

**Stage 5:** Explicit and reflective discussion

Individual students then …

**Stage 6:** Write an investigation report

The report then goes through a …

**Stage 7:** Double-blind group peer review

Each student then …

**Stage 8:** Revises and submits his or her report

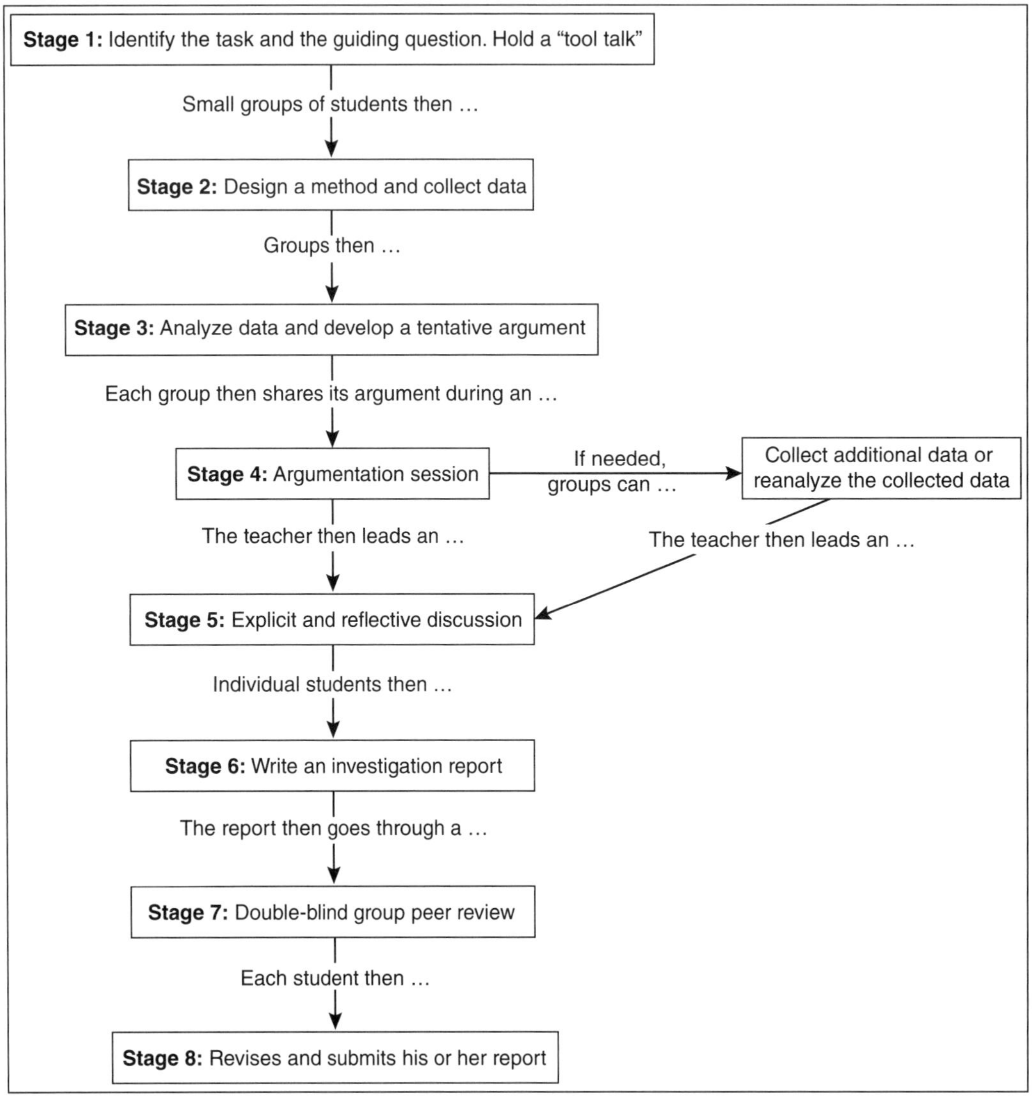

*Source:* Sampson et al. 2014.

This instructional model reflects current research about how students best learn science (Bransford, Brown, and Cocking 2000) and is also based on research on how to engage students in the scientific practice of argumentation, as well as the other practices of science. The role of the teacher in ADI is quite different from the teacher's role in "typical" laboratory experiences. Table 7.3 shows the difference between laboratory instruction using the ADI model and laboratory instruction that is not consistent with the ADI model.

## Table 7.3. Teacher Behaviors During the Stages of the ADI Instructional Model

| Stage | Teacher Behaviors | |
|---|---|---|
| | **Consistent With ADI Model** | **Inconsistent With ADI Model** |
| 1: Identification of the task and the guiding question; "tool talk" | • Sparks students' curiosity<br>• "Creates a need" for students to design and carry out an investigation<br>• Organizes students into collaborative groups<br>• Supplies students with the materials they will need<br>• Holds a "tool talk" to show students how to use equipment and/or to illustrate proper techniques<br>• Provides students with hints | • Does not have students read the lab handout<br>• Tells students that there is one correct answer<br>• Tells students what they "should expect to see" or what results "they should get" |
| 2: Designing a method and collecting data | • Encourages students to ask questions as they design their investigations<br>• Asks groups questions about their method (e.g., "Why do you want to do it this way?") and the type of data they expect from that design<br>• Reminds students of the importance of specificity when completing their investigation proposal | • Gives students a procedure to follow<br>• Does not question students about their method or the type of data they expect to collect<br>• Approves vague or incomplete investigation proposals |
| 3: Data analysis and development of a tentative argument | • Reminds students of the research question and the components of a scientific argument<br>• Requires students to generate an argument that provides and supports a claim with genuine evidence (data + an analysis of the data + an interpretation of the analysis)<br>• Asks students what opposing ideas or rebuttals they might anticipate<br>• Encourages students to justify their evidence with scientific concepts | • Requires only one student to be prepared to discuss the argument<br>• Moves to groups to check on progress without asking students questions about why they are doing what they are doing<br>• Does not interact with students (uses the time to catch up on other responsibilities)<br>• Tells students that their claim is right |

## Table 7.3. (*continued*)

| Stage | Teacher Behaviors | |
|---|---|---|
| | **Consistent With ADI Model** | **Inconsistent With ADI Model** |
| 4: Argumentation session | • Reminds students of appropriate behaviors during discussions<br>• Reminds students to critique ideas, not people<br>• Encourages students to ask peers questions<br>• Keeps the discussion focused on the elements of the argument<br>• Encourages students to use appropriate criteria for determining what does and does not count | • Allows students to criticize or tease each other<br>• Asks questions about students' claims before other students can ask<br>• Allows students to use inappropriate criteria for determining what does and does not count |
| 5: Explicit and reflective discussion | • Discusses the content at the heart of the investigation and important theories, laws, or principles that students can use to justify their evidence when writing their investigation reports<br>• Explains one or two crosscutting concepts using what happened during the lab investigation as an example<br>• Highlights one or two aspects of the nature of science and/or scientific inquiry using what happened during the lab investigation as examples<br>• Encourages students to identify strengths and limitations of their investigations<br>• Discusses ways that students could improve future investigations | • Provides a lecture on the content<br>• Skips over the discussion about the nature of science and the nature of scientific inquiry in order to save time<br>• Tell students "what they should have learned" or "this is what you all should have figured out" |
| 6: Writing the investigation report | • Reminds students about the audience, topic, and purpose of the report<br>• Provides the peer-review guide in advance<br>• Provide examples of a high-quality report and an unacceptable report<br>• Takes time to write the report in class in order to scaffold the process of writing each section of the report | • Has students write only a portion of the report<br>• Allows students to write the report as a group<br>• Does not require students to write the report, in order to save time |

**Table 7.3. (*continued*)**

| Stage | Teacher Behaviors | |
|---|---|---|
| | **Consistent With ADI Model** | **Inconsistent With ADI Model** |
| 7: Double-blind peer group review | • Reminds students of appropriate behaviors for the peer-review process<br>• Ensures that all groups are giving a quality and fair peer review to the best of their ability<br>• Encourages students to remember that while grammar and punctuation are important, the main goal is an acceptable scientific claim with supporting evidence and a justification of the evidence<br>• Ensures that students provide genuine feedback to the author when they identify a weakness or an omission<br>• Holds the reviewers accountable | • Allows students to make critical comments about the author (e.g., "This person is stupid") rather than their work (e.g., "This claim needs to be supported by evidence")<br>• Allows students to just check off "Yes" on each item without providing a critical evaluation of the report |
| 8: Revision and submission of the investigation report | • Requires students to edit their reports based on the reviewers' comments<br>• Requires students to respond to the reviewers' ratings and comments<br>• Has students complete the checkout questions after they have turned in their report | • Allows students to turn in a report without a completed peer-review guide<br>• Allows students to turn in a report without revising it first |

Used with permission from Sampson et al. 2014.

So there we have a short list of instructional models developed over time by teachers, curriculum developers, and researchers, including a current research-based model that explicitly supports the practices of science. And now we will examine instructional strategies that support conceptual understanding and can be used within the instructional models discussed in this chapter.

## Questions for Personal Reflection or Group Discussion

1. What do you think is the benefit of using an instructional model to design instruction?
2. What type of instructional model do you use to design instruction or is in your curriculum materials? How does it compare with the models described in the chapter?

3. What similarities between learning cycle frameworks do you notice in Table 7.1 on page 141? What differences do you notice?
4. Read Dr. Arthur Eisenkraft's article "Expanding the 5E Model," in the September 2003 issue of *The Science Teacher.* How does this article further support the need to explicitly include elicitation and extending in the original 5E Model?
5. Read the *Science Scope* article "Conceptual Change in the Classroom" (Mascazine and McCann 1999). How does this article further your understanding of the Conceptual Change Model (CCM)?
6. What do you see as a major difference between the 5E Model and the CCM Model? When would you use the 5E instead of the CCM or vice versa?
7. Of the four features listed that characterize conceptual change teaching, which of those features are most prevalent in your classroom? Which would you like to work on more?
8. How does the argument-driven inquiry (ADI) model differ from other instructional models? How do you think this model can be modified for younger children?
9. Based on what you have learned from Chapters 1–7 in this book, why do you think a chapter on instructional models was included in this book? What is the connection between instructional models and teaching for conceptual change?
10. Choose one "golden line" from this chapter (a sentence that really speaks to or resonates with you). Write this on a sentence strip and share it with others. Explain why you chose it.
11. What was the biggest "takeaway" from this chapter for you? What will you do or think about differently as a result?

## Extending Your Learning With NSTA Resources

1. Read the guest editorial by Rodger Bybee, "The BSCS 5E Instructional Model: Personal Reflections and Contemporary Implications," in the April/May 2014 issue of *Science and Children*. How does the 5E Model fit with current science education movements and the *NGSS*?
2. Go to the NSTA journals search page on the NSTA Learning Center at *http://learningcenter.nsta.org/products/journals.aspx*. Search for articles that focus on or use the learning cycle (search term: learning cycle). Download and read one of the articles. How was the learning cycle described and used in the article?
3. Go to the NSTA journals search page on the NSTA Learning Center at *http://learningcenter.nsta.org/products/journals.aspx*. Search for articles that focus on or use the 5E Model (search word: 5E). Download and read one of the articles. How was the 5E Model described and used in the article?

4. Elementary teachers combine children's trade books with inquiry science in NSTA's *Picture Perfect Science* series (Ansberry and Morgan 2007, 2010; Morgan and Ansberry 2013). Examine ways the authors use the 5E Model to support inquiry in designing K–4 science lessons.

5. Analogies are ways to make abstract concepts more understandable to students. Read and discuss this article on how analogies can be used throughout the 5E Instructional Model (example is from a high school chemistry lesson): Orgill, M., and M. Thomas. 2007. Analogies and the 5E Model. *The Science Teacher* 74 (1): 40–45.

6. Check out the Community Forums on the NSTA Learning Center site at *http://learningcenter.nsta.org/discuss*. There are several discussions centered on use of the 5E Model you can participate in or start a discussion thread of your own.

7. Read and discuss the chapter, "From Wyoming to Florida, They Ask, 'Why Wasn't I Taught this Way?'" In *Inquiry: The Key to Exemplary Science* (Yager 2009). This chapter discusses how teachers are discovering and using the Conceptual Change Model.

8. Read and discuss the description of the argument-driven inquiry model in the NSTA Press books: *Argument-Driven Inquiry in Biology: Lab Investigations for Grades 9–12* or *Argument-Driven Inquiry in Chemistry: Lab Investigations for Grades 9–12*. Examine the sample lab investigations. How are they similar to or different from the way you currently include lab investigations in your instruction?

# Chapter 8

## What Are Some Instructional Strategies That Support Conceptual Understanding?

In Chapter 7, we described the use of instructional models for teaching science for conceptual understanding. Instructional models provide a framework for organizing instruction. In this chapter we will describe strategies that can be used to promote learning within an instructional model. There are many strategies that researchers and practitioners have determined to be effective in supporting conceptual understanding. We cannot possibly mention them all. We have selected 15 strategies that we have used in our work with students and teachers. These strategies are designed to help you build a bridge between where your students are in their initial thinking (e.g., naive ideas, partially formed ideas, preconceptions from everyday experience) and where they need to be to understand and be able to use scientific concepts. Several of these strategies are considered conceptual change strategies.

The strategies, which we have organized alphabetically, can be integrated into an instructional model and used within a variety of contexts, laboratory experiences, and instructional materials. Examine the strategies and choose one to try out with your students or with the teachers you work with. Once you become comfortable with and versatile in using the strategy, try another, eventually building your repertoire. The 15 strategies we will explore in this chapter are as follows:

1. ABC-CBV (Activity Before Concept, Concept Before Vocabulary)
2. Analogies and Metaphors
3. Argumentation
4. Claim-Support-Question (CSQ)
5. Concept Mapping
6. Discrepant Questioning Used With Visual Models
7. Group Interactive Frayer Model
8. KWL

9. Our Best Thinking Until Now
10. Predict-Observe-Explain Sequences
11. RECAST Activities
12. Role Playing
13. Talk Moves
14. Thinking About Thinking: Supporting Metacognition
15. Thought Experiments

## Author Vignette

It was a hot, humid summer evening in July. I went out to dinner with a group of teachers during a summer institute in Maine. As we sat outside sipping cold beverages, I looked at the teachers and whispered, "Remember how we talked about how misconceptions about basic phenomena can follow us right into adulthood if they have never been challenged? Well, watch this!" Our server, a young college student, came over and I proceeded to hold up my glass and point at the wetness on the outside of the glass. I asked our server, "My friends and I are wondering why the outside of our glasses is all wet?" I predicted exactly what the server would say, and lo and behold, I was right. He immediately responded, "Oh, that's just condensation." I then asked, "But we can't figure out how the condensation got on the outside of the glass." He proceeded to explain to us, by pointing at the liquid inside the glass that the "coldness" from our drinks comes out and around, touches the outside of the glass, and forms condensation. I thanked him, while surreptitiously winking at the teachers who had a wide-eyed look of shock on their faces. I have asked this question several times of adults, particularly when I am sipping cold beverages on hot summer days, and predictably, they often reply using the term *condensation*, yet their explanation reveals a lack of conceptual understanding of the word and the concept. It's a real-life example of how students and adults will often use scientific terminology without understanding the concept!

—Page Keeley

## Strategy #1: ABC-CBV (Activity Before Concept, Concept Before Vocabulary)

Former NSTA president, Distinguished Professor of Science Education at the University of Massachusetts Boston, and highly accomplished former high school teacher Dr. Arthur Eisenkraft coined the phrase "activity before concept," or ABC teaching for short. *Active Physics* and *Active Chemistry*, two science curriculum programs written by Dr. Eisenkraft, are examples of exemplary instructional materials that use the ABC approach. ABC teaching is the opposite of the way lessons are often presented in instructional materials or teacher-designed activities. Instead of presenting and explaining a concept and then launching into activities and investigations, ABC teaching begins with the activity or investigation so students first have an opportunity to interact directly with phenomena, objects, or materials before being presented with the concept. ABC can be extended with CBV (concept before vocabulary). After students have been presented with the concept, which they can attach to the activity they experienced, the vocabulary is then introduced.

Typical lab experiences often follow presentation of concepts and essentially become verification or procedural experiences. Even though these activities are designed to reinforce the concepts and vocabulary that have been presented to students, if students' initial thinking about a phenomenon or concept has not been engaged, challenged, and worked through during an activity, lab, or discussion, the experiences often do little to develop conceptual understanding. Students need to have observations or rich experiences for discussion and argumentation upon which to "hang" the concept or vocabulary to conceptually understand it. The following is an example of ABC-CBV teaching related to the concept of *transfer of energy during a phase change*:

The teacher uses the formative assessment probe "Ice Water" (Figure 8.1) to elicit students' initial ideas about the phenomenon of phase transition, connecting it to the concept of *transfer of energy*. Many students will predict that the temperature of the "ice water" will decrease after more ice cubes are added. This makes sense to them since their experiences adding ice to water are usually to "make the water colder." After discussing their ideas, students then test their predictions and find out that the temperature of the ice water is the same regardless of how many more ice cubes they add. This creates a

**Figure 8.1. "Ice Water" Probe**

**Ice Water**

Christine put five ice cubes in a glass. After 20 minutes, most of the ice had melted to form "ice water." There were still some small pieces of ice floating in the water. Christine measured the temperature of the ice water then added five more ice cubes to the glass. She measured the temperature three minutes later. What do you predict happened to the temperature of the "ice water" three minutes after she added more ice?

**A** The temperature of the "ice water" increased.

**B** The temperature of the "ice water" decreased.

**C** The temperature of the "ice water" stayed the same.

Circle the answer that best matches your thinking. Explain what happens to the temperature of "ice water" when more ice is added.

*Source:* Keeley and Tugel 2009.

disequilibrium that opens up the opportunity for the teacher to explain what is happening using the concepts of *heat* and *transfer of energy*. The teacher explains that heat is transferred as the solid ice cubes absorb energy (and introduces the term *thermal energy*) from the surrounding air in the room, and that the temperature remains constant until all of the ice melts. It does not get colder or warmer as the ice melts and two phases are present. Once all the ice melts, the energy in the system will result in increased motion of the liquid molecules and the temperature will increase (the teacher makes sure students have the opportunity to observe and measure this). The teacher then connects three often-confused terms—*thermal energy, heat,* and *temperature*—to the concept of *energy transfer during a phase transition* (Keeley and Tugel 2009).

The *Everyday Science Mysteries* series uses a similar puzzling story to engage elementary and middle school students in developing hypotheses, testing them, and analyzing their data to complete the story. The concepts and terminology are not provided prior to the activity, but rather are integrated into the story and explicitly brought to the forefront after students have had opportunities to surface their ideas, discuss them, test them, and revise them after examining and making sense of the data they collect. The following mystery story, "How Cold Is Cold?" can also be used to develop the concept of transfer of energy and connect it to the formal terminology of *heat, temperature,* and *thermal energy* (Konicek-Moran 2013c).

## How Cold Is Cold?

Kristin filled her glass with ice cubes from the freezer, all the way up to the top! She then filled the glass with lemonade and sat down to drink it. The day was hot and muggy and Kristin did not take long to finish her drink. When she was finished, she dumped almost a full glass of ice cubes into the sink.

Kristin's father had been watching the entire scene. "You know, Krissy," he said," you don't have to waste all of that ice. Why do you put so much ice into your glass?"

"I like my lemonade really cold," she responded, "and the more ice I put in, the colder the lemonade gets."

"Are you sure about that?" asked her dad.

"Of course," answered Kristin. "It makes sense. More cold ice makes a cold drink, well ... colder."

"Maybe," said her dad, "more ice might make it cool down faster, but would it really make it colder? Look! You threw away almost all of the ice!"

"It was cold enough, so I drank it all down. I can't help it if all of the ice didn't melt. Besides, if I let all of the ice melt, the lemonade would have gotten colder and colder and maybe too cold to drink. There was a lot of cold in the ice that had to go into the drink and the more ice, the more cold there was to cool the drink."

"I don't know," mumbled her dad. "Something doesn't quite make sense here. Could the lemonade get colder than the ice that's in it?"

"Well, I think so," mumbled Kristin cautiously. "Or maybe not. I don't really know. More ice would keep on making it colder as long as there was still ice, wouldn't it?"

"We need to do some experimenting," said her dad. "We need an hypothesis or two. It looks like we have a least a couple of questions here."

*Source:* Konicek-Moran 2013c.

Many teachers expect their students to learn science vocabulary, or the technical terms of science, in order to read about and understand scientific concepts (Glen and Dotger 2009). Science has a tremendous amount of technical terminology, and vocabulary and definitions are often given precedence in many textbooks and instructional materials. The ABC-CBV approach allows students to experience a situation first and then connect a concept and the scientific terminology associated with the concept to the thought-provoking experience, thus deepening their conceptual understanding.

## Strategy #2: Analogies and Metaphors

The late Bob Samples, educator extraordinaire, once said that if you could not find a metaphor for a concept, you probably didn't really understand it (1974, personal conversation). Bob spent a great deal of time and energy working with children and creating curricula that focused on metaphor, analogies, creative uses of everyday materials, and the arts. He liked to tell about the time he gave an inner-city class some Polaroid cameras and asked them to photograph "power" (this was before the days of digital cameras). Some came back with the usual pictures of power lines and big cars, but many also brought pictures of neighborhood pimps, drug dealers, police, and even the school principal. The pictures allowed him to peek inside his students' minds and see how a simple term like *power* had so many different connotations to his students.

The Synectics Group of Cambridge, Massachusetts, a problem-solving and creativity consulting group started by George Prince and William Gordon, coined an interesting metaphor for helping students learn: "making the strange familiar" (Prince 1970; Gordon 1961). Think of a new concept as being strange to a student. Connecting it to something with which he or she is familiar will help to align the new idea with a familiar one.

Take for example the theory of blood circulation developed by William Harvey in Elizabethan England. At that time, it was the belief that blood circulated through the body by way of contractions of the arteries, perhaps because physicians were aware of the pulse in various parts of the body. Harvey theorized that the blood in the body moved from the heart through arteries and returned via veins. Besides his experience with dissections, he was led by a strong religious belief that nature mirrored the motion of the planets, a circular path. The Sun was the center of the universe, so then everything in the universe would mimic the motion of the planets. He concluded that the heart was the beginning of life, and, like the Sun, is the center of the body's system. So, following the analogy of the planets, circulation was the answer. Harvey was never able to find the capillaries that formed the junction between the veins and arteries and completed the cycle. His contraction and circulation theory, however, was shown to be correct later when microscopes and other magnifiers became available.

If a teacher were to need an analogy for the heart, a pump would be a likely choice. What a perfect way to help students understand this alien concept: The "familiar" idea of a pump, by means of a metaphor, connects with the "strange" idea of the heart's function.

The information already understood about the pump is mapped to the heart and thus the transfer of learning is accomplished. One must be careful, however, to point out that analogies and metaphors have shortcomings. They are conceptual models, and like all models, have limitations. There are some parts where direct comparison may not hold true. The base serves only as a conduit between the known and the unknown.

John Clement of the University of Massachusetts Amherst has done a great deal of research on this topic. In his book *Creative Model Construction in Scientists and Students: The Role of Imagery, Analogy, and Mental Simulation* (2009), he tells of his study of professional scientists and novices and their use of analogy and imagery to solve physics problems. He describes the work of Gentner (1983) and Forbus, Gentner, Everett, and Wu (1997) who developed a theory on how the process of analogical reasoning works:

- The analogous case is accessed by being activated associatively and retrieved from permanent memory.
- A mapping is generated between corresponding entities in the base and the target and the soundness of the analogy is assessed.
- One or more key elements are inferred in the target. (Clement 2009, p. 23)

In the famous "book on the table" problem for high school physics students, Clement and his colleagues studied how students came to realize that a table supporting a book actually provided "an equal and upward force on the book." This is an entirely counterintuitive concept since most students think of a table as completely inanimate. It is thought of as mainly a barrier to the gravity that is exerting a downward force on the book. How could a passive object exert an upward force?

Clement's group added what they called bridging analogies, such as having the student hold the book and feel the upward pressure needed to keep it from falling, introducing a spring between the table and book to show that the spring puts an upward force on the book, and finally resting the book on a thin piece of board that visually demonstrates a distortion. These activities, analogies of the forces involved, allowed the students finally to understand the concept (Clement et al. 1987). Finally, Clement makes the following observation about changing students' concepts using these techniques:

> *Such strategies would take advantage of positive elements of prior knowledge by building on students' existing valid physical intuitions. It is perhaps confusing that we are attempting to build on students' conceptions in order to change their conceptions. However there is evidence that students' have both useful and competing intuitive conceptions (from the perspective of the scientific theory being taught). (2009, p. 39)*

This, of course, corroborates our position about listening to students' ideas carefully and moving on from there. Clement's group believes that students' naive ideas are not only a good starting point but are necessary places for beginning the development of movement toward "targets" that are synonymous with the targets (core ideas) developed in the *NGSS*.

There is a vignette in *Ready, Set, SCIENCE!* showing how teacher Richard Sohmer introduced the analogy between puppies and molecules to show students how pushing forces explained the ideal gas laws. Sohmer asked the students to substitute, mentally, puppies for molecules, place them in enclosed spaces with moveable walls and then change the variables so the puppies became either more numerous or more excited. The resulting movement of the walls due to the force of the "bounding" puppies created the concept of a force as a push and eliminated the explanation of "sucking" to explain phenomena that involved air pressure from molecules (Michaels, Shouse, and Schweingruber 2008). Even though Sohmer made sure that there were graphics depicting the puppies analogy, it was important that the students had the opportunity to internalize the mental model so they had it firmly implanted in their memory. They did this through discussion and interaction with the teacher and each other. Without such a mental model of bouncing molecules transferring energy to their surroundings, the analogy would not be viable for them in other situations (e.g., understanding heat flow, the ideal gas laws, and other concepts involving the behavior of molecules and atoms).

## Strategy #3: Argumentation

"Science learning can be very effective when it is grounded in a task that supports multiple predictions, explanations, or positions. In such a setting, children have reasons to 'argue' (to agree and disagree) and to back up their positions with evidence" (Michaels, Shouse, and Schweingruber 2008, p. 68). Unfortunately many science textbooks and instructional materials present science as unequivocal and uncontested, essentially encouraging students to accept what is presented without question. Providing students with opportunities to both construct arguments and to critically analyze arguments contributes to their understanding of the role of argument in science while simultaneously challenging and deepening their conceptual knowledge. In order to process, make sense of, and learn from their ideas, observations, and experiences, students must talk about them (Michaels, Shouse, and Schweingruber 2008). Argumentation is a form of science talk that has important implications for conceptual teaching and learning in science.

In the past decade, many studies have been conducted on the importance of argumentation in science. In the relationship between argumentation and conceptual understanding, it is not entirely clear which one is the contributor to the other (Venville and Dawson 2010), but studies do show that engaging students in argumentation does positively influence their understanding of a topic (Albe 2008; Bell and Linn 2000).

Unlike the everyday types of arguments students engage in with their friends or family, the goal of argumentation in the science classroom is to promote an understanding of a phenomenon or situation and to persuade others of the validity of one's ideas. Students do not argue to "win" in the science classroom but instead argue to share, make sense of, and learn about ideas.

To engage in argument, students must first have an interesting question or be presented with a puzzling phenomenon or situation to argue about. The *Uncovering Student Ideas in Science* and the *Everyday Science Mysteries* series provide a source of interesting questions and puzzling situations to engage students in argumentation. Constructing an argument requires that students use evidence and consider counterarguments. As an instructional strategy, argumentation can be facilitated through classroom discussions or linked to writing in the content area. While listening to students present their arguments or reading students' written arguments, the teacher is also gaining valuable insight into students' thinking.

## Author Vignette

I recently worked with a group of K–12 teacher leaders on strategies that develop and support argumentation using the *Uncovering Student Ideas in Science* probes. I modeled the use of the strategy VDR (Vote, Discuss, and Revote) as a way to promote student thinking through argumentation while monitoring for conceptual change. All of the teachers agreed that at some point in their own K–12 or college education, they learned about electric charge; some even taught the concept, so everyone entered the activity with some prior knowledge.

### Does the Example Provide Evidence?

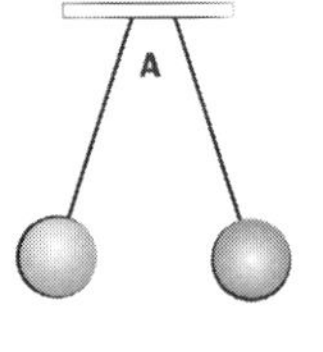


Students in Mr. Miller's class watched a demonstration on electric charge. In the first example of an interaction, Mr. Miller brought two plastic balls together. The balls moved apart when they got near each other, like this (A). In the second example of an interaction, the balls moved toward each other, like this (B). Mr. Miller asked the students if the two examples of interactions provided convincing evidence that both balls in each example were electrically charged. Here is what some students said:

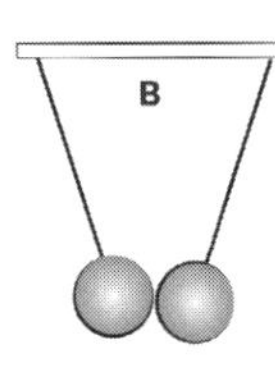


**Faith:** I think both examples provide evidence that all of the balls in both interactions were electrically charged.

**Milo:** I think only the first example provides evidence that both balls in an interaction were electrically charged.

**Judd:** I think only the second example provides evidence that both balls in an interaction were electrically charged.

**Fran:** I think neither example provides evidence that both balls in an interaction were electrically charged.

With whom do you agree the most? ____________ Explain why you agree.

*Source:* Keeley and Harrington 2014.

The teachers were presented with the formative assessment probe, "Does the Example Provide the Evidence?" (Keeley and Harrington 2014). After ensuring that the context of the probe situation was clear, teachers were asked to complete the probe individually by selecting the person they most agree with and explaining why they agree more with one person than the others. This was their first step in examining the evidence in preparing to construct an argument to defend their thinking. We then took a vote (first step in the VDR strategy) and recorded on a chart how many people agreed with each of the answer choices. The majority of the group at this point selected Faith as the person they most agreed with. The chart was posted in the room so everyone could track the group's thinking throughout the stages of the VDR.

The next step in the VDR strategy involved discussion in pairs. During this stage the teachers could only talk to one other person. In pairs, they shared their arguments for why they agreed with the person they selected. If they selected the same person, together they discussed and solidified their ideas and constructed an argument they would both use to support their answer. If their answer choice differed, they engaged in discussion sharing their reasons for selecting one person over another and in the process constructed and strengthened their own arguments. After allowing enough time for the teachers to develop and share their arguments, we took a revote and recorded the results of the pair discussion. We noticed that no one now argued for Judd's idea and that one less person argued for Fran. There were now five people who argued for Milo, but the majority still leaned toward Faith.

The next round of the VDR involved discussion in small groups. Since I knew that Milo was the best answer, and I was able to observe during the voting who picked Milo, I made sure to distribute the teachers who selected Milo among the small groups so that most of the groups would have at least one person one could argue for Milo. We formed groups of five and the teachers were given five minutes to engage in argumentation with the purpose of convincing others of the validity of their ideas. During the small-group discussion, I was able to circulate and listen to their arguments, noting individuals who could present a strong argument for Milo, as well as for Faith and Fran during the next round of VDR, the whole-group discussion. After the small groups presented and evaluated each other's arguments, a revote was taken. Now there was a significant shift toward Milo. Only two teachers held onto their argument for supporting Fran, and there were still a dozen teachers who felt the argument for Faith's idea was the most compelling.

We now moved into the final phase of the VDR—the whole-class discussion. We opened up the argumentation session to the idea that got the fewest number of votes, Fran. One of the teachers who selected Fran and held on to her idea throughout the previous arguments said the evidence was inconclusive because you didn't know what the balls were made of. Another teacher offered a rebuttal by saying that the material did not matter. The fact was that the balls moved apart in the first interaction, and moved together in the second interaction. Since the balls were not magnets, it had to be because they were both electrically charged.

Those who were still going with Faith presented their arguments, which were primarily based on their prior knowledge that like charges repel, which is why both of the balls in the first interaction moved apart; and that unlike charges attract, which explains why they moved together. Because the balls moved together in the second interaction, one of the balls had to be positive and the other had to be negative.

Those who chose Milo offered a rebuttal of the Faith argument. They agreed that both of the balls in the first interaction were electrically charged. They both had to be positively charged or they both had to be negatively charged. However, in the second interaction one of the balls could be uncharged (neutral) and there would still be an attraction between a neutral and a charged object. They supported their argument with the example of rubbing a balloon on your hair and sticking it to a wall or rubbing a plastic comb and picking up bits of paper. Neither the wall nor the paper was electrically charged, but there was still an interaction between the objects. Therefore Milo's idea is the best one, as the only convincing evidence that both objects are electrically charged is in the first interaction.

I have only captured snippets of the arguments here. There were many more ideas and concepts that were exchanged during each of the argumentation sessions. We took a final revote and the results were: Faith–2, Milo–29, Judd–0, and Fran–0.

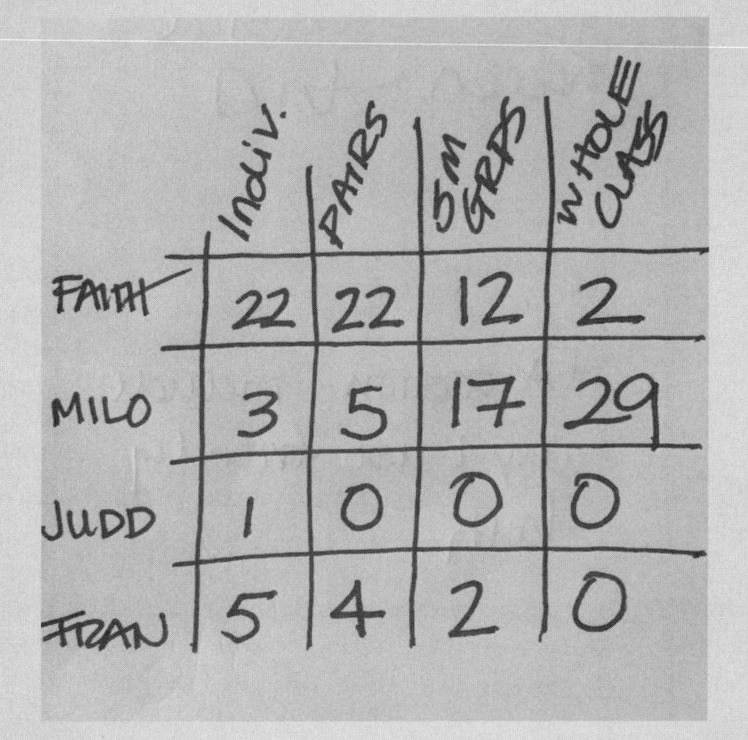

| | INDIV. | PAIRS | SM GRPS | WHOLE CLASS |
|---|---|---|---|---|
| FAITH | 22 | 22 | 12 | 2 |
| MILO | 3 | 5 | 17 | 29 |
| JUDD | 1 | 0 | 0 | 0 |
| FRAN | 5 | 4 | 2 | 0 |

Even though there were still two holdouts for Faith, the arguments of their peers were convincing enough to enable others to change their thinking as they evaluated information that they had not considered previously. After experiencing it themselves, the teachers agreed that argumentation is a powerful way to help students reconstruct or refine their ideas by examining the evidence and arguments presented by others. At the same time students whose initial ideas were scientifically correct were able to solidify them by strengthening their arguments. What was most surprising about this strategy for the teachers was that

> the instructor did not do any direct teaching of the concepts, but rather facilitated the process and monitored how ideas were changing and developing as the class moved toward understanding the interaction between charged and uncharged objects. We agreed that argumentation using the VDR strategy is an example of "teaching with your mouth shut" and that too often teachers feel they have to be the givers of information, when in fact, if there are students in a group that have some understanding of the concepts, they can be powerful initiators of conceptual change for other students.
>
> —Page Keeley

Students can also construct written arguments in response to a question or a puzzling situation that requires the use of evidence and the consideration of counterarguments. For example, the formative assessment probe "Is Earth Really 'Round?'" (Keeley and Sneider 2012) confronts students with several different arguments. After students have had an opportunity to discuss the evidence for the validity of each of the arguments presented in the probe, they can construct a final written argument that is supported with evidence and considers the validity of the alternative arguments. A framework that supports this type of argumentation writing that includes the reasons for one position versus the opposing views was developed by two researchers widely known for their work in connecting science and language literacy, and is described as follows, using the "Is Earth Really 'Round?'" probe (Figure 8.2) as the basis for the written argument (Wellington and Osborne 2001, p. 74).

## Figure 8.2. "Is Earth Really 'Round'?" Probe

# Is Earth Really "Round"?

Five friends were talking about the shape of the Earth. They each agreed the Earth is round. However, they disagreed about what "round" really means. Here are their ideas about a "round" Earth:

**Chuck:** "I read somewhere that Columbus or Magellan or someone proved the Earth is round like a round island. He sailed all around the island, and came back to the same port."

**Sara:** "I know the Earth doesn't look round. That's just because we live in a flat area. Other people can see it's round because they live near mountains and hills."

**Takesha:** "'Round' means that the whole Earth is shaped like a ball. It just looks flat because we can only see a small part of the ball."

**Arnold:** "You're right that 'round' means 'round like a ball,' but it looks flat because we live on the flat part in the middle. The upper part of the ball is the sky, and the bottom part is the solid Earth, where people live."

**Missy:** "Everyone knows that the round Earth is a planet in the solar system, like Mars and Jupiter. People get mixed up because 'earth' is also another name for the ground."

Who do you think has the best explanation? ____________ Explain why you think it is the best explanation, and use a drawing to support your explanation.

*Source:* Keeley and Sneider 2012.

### *A Framework to Support a Written Argument*

There is a lot of discussion about whether …

[The Earth is flat.]

The people who agree with this idea claim that …

[It looks flat.]

[If it were round, people would fall off, ships would drop off the edge.]

They also argue that …

[Photographs could be faked.]

A further point they make is …

[It looks flat in photographs taken from the air.]

However, there are also strong arguments with evidence against this view. These are …

[Pictures of the Earth from space showed that it is round.]

[The shadow of the Earth on the Moon is round.]

[It explains why shadows of the Sun vary from nothing at the equator to much longer lengths toward the North Pole.]

Furthermore they claim that …

[People are held onto a round Earth by the force of gravity.]

After looking at different points of view, and the evidence, I think that …

Argument is an important feature of science and by building argumentation strategies into your instruction, you not only foster students' ability to think scientifically and reason from evidence, you also build a conceptual understanding of how scientists communicate through writing and through discussions with other scientists.

## Strategy #4: Claim-Support-Question (CSQ)

CSQ is a thinking strategy designed to help students identify claims and hold them up to thoughtful scrutiny (Richhart, Church, and Morrison 2011). Students first think about making claims by looking for patterns, generalizations, prior observations, and scientific concepts they can use to explain situations or phenomena. For example, before launching into a unit topic on magnetic interactions, the teacher might ask students, "What claims can you make about how magnets interact with objects?" Or, after students have had opportunities to explore magnetic interactions, the teacher might ask, "What claims can you make about magnetic interactions based on your observations?" Students might generate claims such as magnets can move objects without touching them, magnets attract metals, or magnets will work in water. As students generate their claims, the teacher lists them on a three-column chart, under *C* for claim, leaving space to add evidence, questions, and more claims.

Once the claims are listed, the teacher asks, "What have we observed or know that will support these claims?" Students provide the evidence for the claims under the *S* column for support. This step helps them consider what constitutes evidence and whether they have sufficient support for the claim. For example, in response to the claim "magnets attract metals" students might add their observation that magnets pick up steel paper clips.

Under the *Q* column, the teacher encourages the students to question the claims and support (evidence for the claims) and think beyond the support for the claims that has already been listed. Students are encouraged to evaluate the evidence for the claim and determine whether it is valid as well as sufficient in supporting the claim. For example, students may decide that metals other than steel need to be tested in order to support the claim that magnets attract metals.

As students share their thinking, they may add additional claims to the chart. Documenting the CSQ on a chart for all to see and examine allows students to build on their thinking as well as challenge others' thinking.

## Strategy #5: Concept Mapping

A concept map graphically connects terms with connecting lines on which are written connection terms. The idea of concept mapping was developed in the 1970s by Joseph Novak at Cornell University (Novak 1998). For example, suppose the topic is *water cycle* and the concept of *evaporation* is part of the map. The concept *evaporation* can be connected to a concept such as *energy* with the connecting line saying "requires an input of." The technique has been used in early childhood (Birbili 2006) and in elementary, middle (Vanides et al. 2005), high schools (Kern and Crippen 2008) and colleges (Preszler 2004). Figure 8.3 shows an example of a concept map made by a middle school student that was used as an initial elicitation about students' concept of matter. The concept map reveals that the student recognizes matter as taking up space (having volume), but it isn't clear whether the student recognizes that matter also has properties of weight and size that can be measured. The student doesn't include the subconcepts of mass or atoms and seems to focus more on properties of matter,

**Figure 8.3. Concept Map Example**

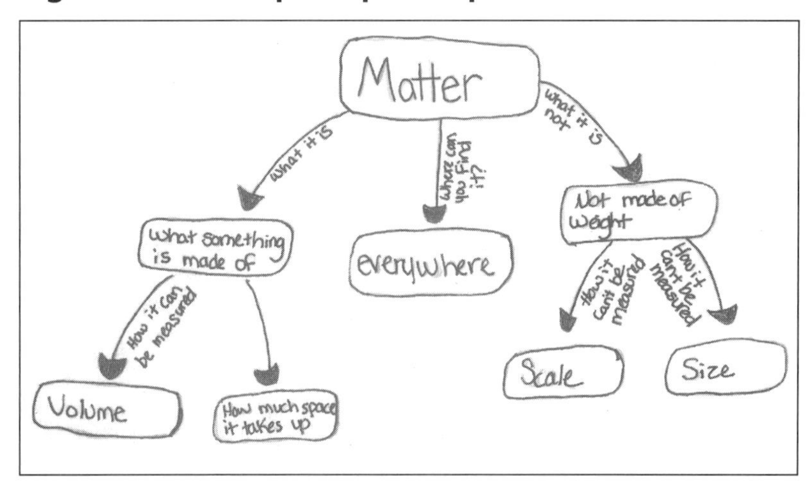

particularly volume. The concept map reveals a limited understanding of the concept of matter. A comprehensive article on the use of concept maps can be found in *Science Scope* titled "Using Concept Maps in the Science Classroom" (Vanides et al. 2005).

Concept mapping can be used as an elicitation prior to instruction or after students have had opportunities to develop ideas. One way to get students to think about the connections among concepts and subconcepts is by giving students a set of concepts and subconcepts printed on cards or sticky notes and asking them to place them around or from a larger concept and to draw lines that connect the concepts and subconcepts with phrases that show the connection. The cards or sticky notes allow students to reposition their concepts and subconcepts as they think through the connections.

Concept maps can be single representations of the way a student thinks about a concept or multiple concept maps can be used to represent students' changing ideas about the concept. This type of concept mapping is called recursive concept mapping. Recursive concept mapping starts with elicitation and involves building on and revising the original concept map as ideas change or are solidified throughout the instructional sequence. Students may revise their maps several times, after discussion and other instructional activities, before ending with a final concept map that demonstrates their conceptual understanding.

*The Science Teacher* article "Mapping for Conceptual Change" (Kern and Crippen 2008) shows a high school example of recursive concept mapping for the concept of *evidence of evolution* in which students use the open source concept mapping software, CmapTools, to complete four iterations of their concept map throughout the instructional sequence. Notice the first map in Figure 8.4 is used at the initial elicitation stage to uncover initial ideas students bring to their learning before instruction.

The teacher has the students complete another map as they begin to develop understanding of the concept, yet some misconceptions are still evident. A third mapping activity reveals the extent to which students are now able to make accurate linkages and which concepts students have not yet developed solid understanding of. The final concept map, as shown in Figure 8.5, is completed at the end of the instructional sequence and is used to provide evidence of a more robust conceptual understanding as well as reflect back on how students' thinking changed.

One of the early pioneers in understanding how children learn, David Ausubel (1963), made a very important distinction between rote learning and meaningful learning. One of the conditions for meaningful learning to take place is that the subject matter to be learned must be conceptually clear and presented with language and examples that relate to the learner's prior knowledge. Concept mapping meets this condition by identifying concepts and linkages held by the learner prior to instruction and then revisiting the concepts and linkages to incorporate new knowledge that helps the learner revise or refine their conceptual understanding.

## Figure 8.4. First Concept Map (Initial Elicitation Stage)

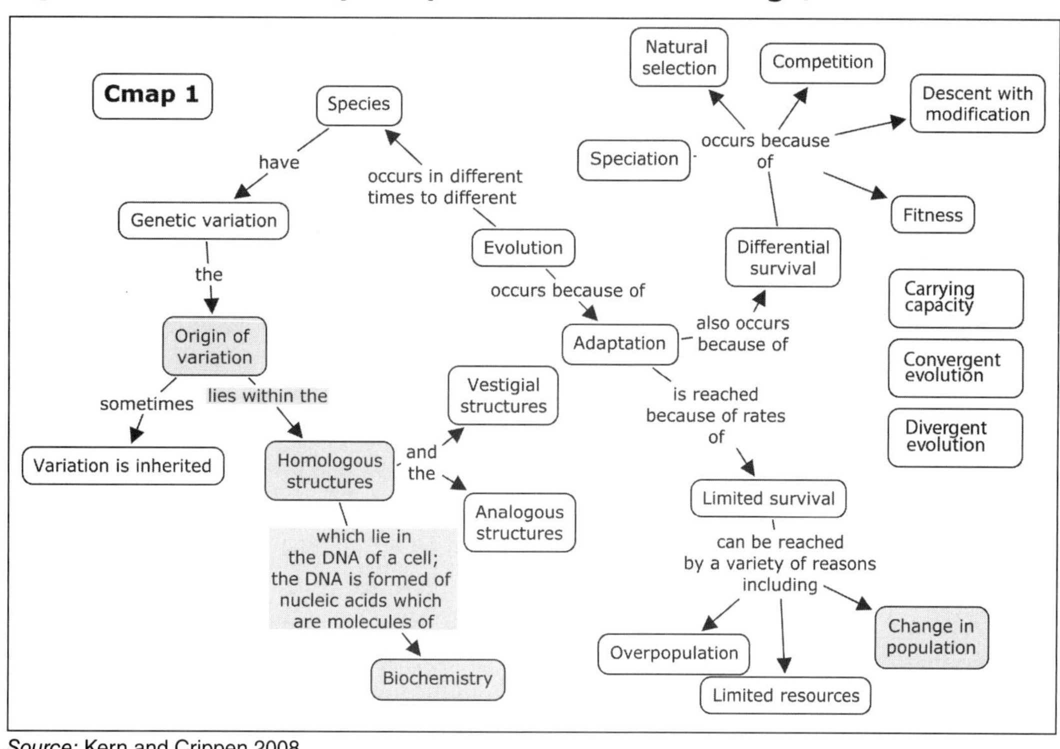


*Source:* Kern and Crippen 2008.

## Figure 8.5. Final Concept Map (Showing Conceptual Understanding)

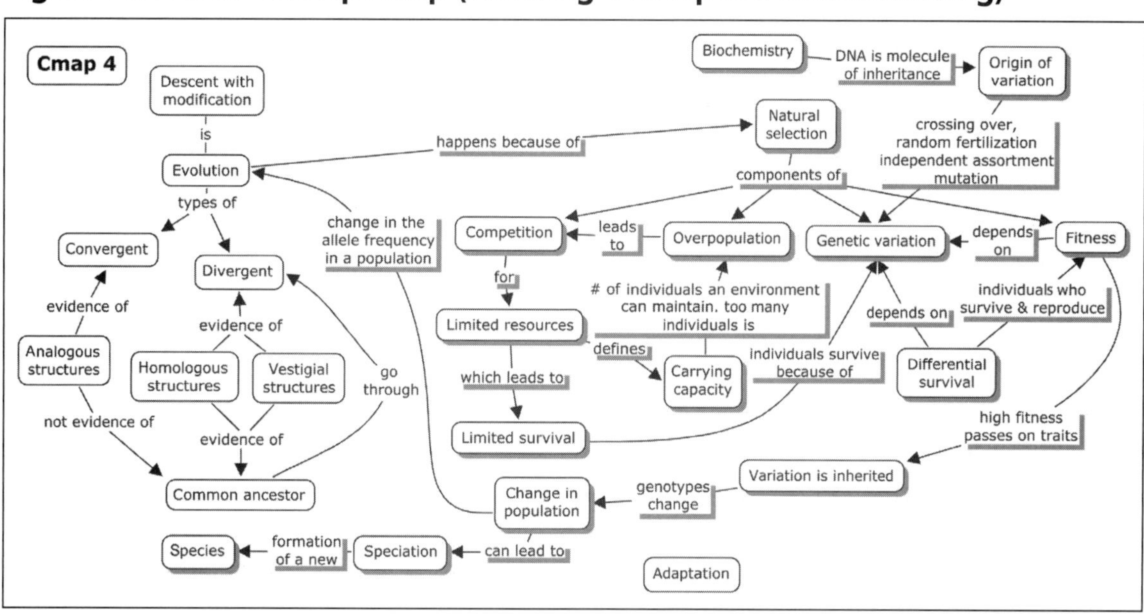


*Source:* Kern and Crippen 2008.

## Strategy #6: Discrepant Questions Used With a Visual Model

This strategy accentuates the importance of the students and the teacher as partners in achieving a target understanding of a particular concept. The teacher's role is to provide scaffolding in the form of discrepant questioning (a term invented by the Clement research group) and guided discussion (Clement et al. 2008). Students are asked to present a model of a concept in a visual manner before any instruction is given. This often results in drawings. If the expression of the concept contains problems, the teacher may ask a "discrepant question" that causes dissonance about their model. The question is geared toward causing the student to rethink the model. Example: The target is how the human heart works. The student draws a heart that is open at the bottom and the teacher asks, "Does your model show blood going out the bottom into the body?" The student then begins to think of the ramifications of such an occurrence and closes the hole in the heart. "Where do we want the blood to go?" asks the teacher. Further revisions follow in small steps and the model evolves into something much closer to the target. Therefore, in this model, concept change occurs in small increments rather than in one major event, with the teacher directing the evolution of ideas. Since the teacher asks for all ideas, the lesson becomes student directed and modified. It takes more time but it allows the students to be in a position to modify their ideas with the help of the teacher and their peers. The articles mentioned show how a teacher uses discrepant questioning and group discussion in a biological domain about the functioning of lungs in a human body. They are trying to develop a model of the lungs that shows how oxygen gets from the lungs to the blood by means of alveoli. This model differs from other strategies that use students' drawings and conceptual models in that it stresses small incremental steps to the target rather than major leaps (Rea-Ramirez 2008; Rea-Ramirez and Nunez-Oviedo 2008).

## Strategy #7: Group Interactive Frayer Model

The Frayer Model was developed by Dorothy Frayer and her colleagues at the University of Wisconsin–Madison (Frayer, Frederick, and Klausmeier 1969). It was originally used as a content literacy strategy to develop vocabulary. In teaching science for conceptual understanding, the Frayer Model is used to elicit and develop students' ideas about a concept. The format is a graphic organizer that organizes prior knowledge about a concept into four sections: an operational definition, characteristics, examples, and non-examples. An operational definition is a definition that is not theoretical but defines a concept through an active process.

## Author Vignette

Picture a classroom of sixth graders, grouped in numbers of four or five. They have a tub of water in front of them and a set of objects including candles, crayons, plastic baskets with holes in them, string, pencils, and the usual array of sink-or-float objects used in classrooms. Our purpose was to try to find out what these students thought about sinkers and floaters and why they belonged in either category.

When the students completed the task of predicting which items would sink or float and then tried out their predictions, we ended up with a group of what the children called, "Yeah, buts." For example, "The bottle lid floats when you put it in gently. "Yeah, but if you just drop it in, it sinks!" The "Yeah, buts" were confusing the students.

I told them a story about a group of students I had heard about who put identical lit candles under identical jars at the identical time and yet came up with great differences in the time it took for the candles to go out. I asked them if they could figure out why there was such a discrepancy. Try as they did, they could not. The answer was, I told them, that some students said the candle was "out" when the flame disappeared. Others said it was out when the wick no longer glowed and still others defined "out" as the time when the smoke stopped rising. ("Where there's smoke, there's fire.") In other words, they had never defined *out*. I then asked them how park officials decided when the ice on a pond was too thin to skate upon. They responded that there was probably a measurement. What then was the definition of *thin ice*? It had to be an operational definition.

They got it! Almost immediately, they began to develop definitions of floaters and sinkers based on operation. They agreed on the operational definition that it was a natural sinker if you placed it on the bottom of the tub and it stayed there; if it came up, it was a natural floater. Progression on the other aspects of the lesson followed over the next few weeks.

—Dick Konicek-Moran

The Group Interactive Frayer Model takes the original Frayer model, which is usually completed as a worksheet, and extends it into a small-group and whole-class activity. The Group Interactive Frayer Model is used to elicit students' prior knowledge and provide an organized structure for them to discuss their ideas with peers in small groups. It gives students an opportunity to initially think about a concept prior to developing a formal understanding of the concept, recognize and consider the range of ideas other groups have about the concept, and revisit their initial ideas after formal instruction in order to revise or refine their thinking. The Group Interactive Frayer Model often provides a bridge between students' operational definition of concepts they encounter in science and the scientific definition of the concept.

**Figure 8.6. "Is It a Rock? (Version 2)" Probe**

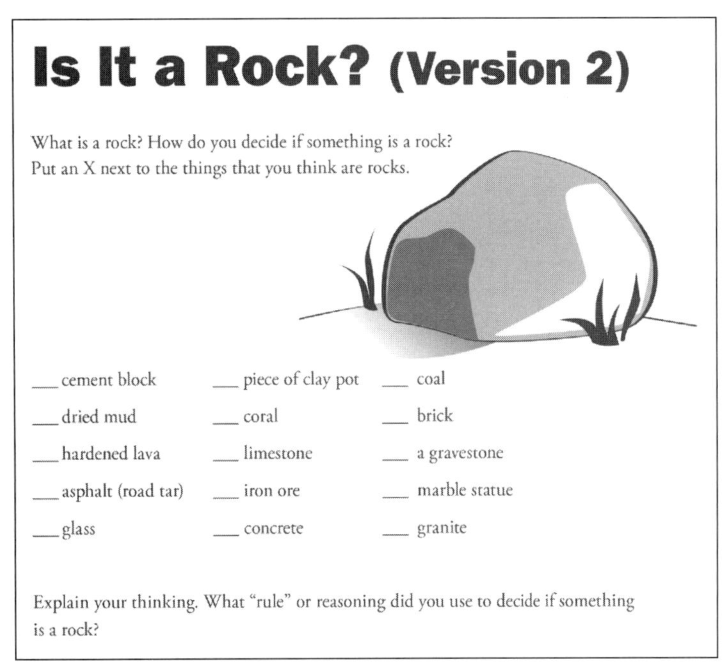

# Is It a Rock? (Version 2)

What is a rock? How do you decide if something is a rock?
Put an X next to the things that you think are rocks.

| | | |
|---|---|---|
| ___ cement block | ___ piece of clay pot | ___ coal |
| ___ dried mud | ___ coral | ___ brick |
| ___ hardened lava | ___ limestone | ___ a gravestone |
| ___ asphalt (road tar) | ___ iron ore | ___ marble statue |
| ___ glass | ___ concrete | ___ granite |

Explain your thinking. What "rule" or reasoning did you use to decide if something is a rock?

*Source:* Keeley, Eberle, and Tugel 2007.

Starting with small groups of three or four students, provide each group a set of four different colored sticky notes or colored paper cut into squares, color-keyed to each of the sections on a Frayer chart. Each small group works together to discuss the concept and record their consensus ideas on each of the sticky notes. For example, the formative assessment probe "Is It a Rock? (Version 2)" (Figure 8.6) was used to elicit students' ideas about things formed geologically versus human-made, rocklike materials (Keeley, Eberle, and Tugel 2007).

Some students may think human-made materials such as concrete, clay pots, glass, and cinder blocks are rocks and fail to recognize that rocks are made through long-term geologic processes. The teacher might decide to use the Group Interactive Frayer Model strategy to have students discuss their responses to the probe, sorting the list of objects into examples of rocks or non-examples of rocks. As they sort and discuss the objects, together students form an operational definition of a *rock* and list characteristics or properties that describe rocks that can be used to distinguish rocks from other materials. They post their sticky notes on the class Frayer chart as shown in Figure 8.7.

The teacher summarizes the ideas on the chart and engages the whole class in a discussion about the concept of a *rock* until they come to consensus as to the best operational definition, the agreed-upon characteristics that distinguish rocks from rocklike materials, and examples and non-examples. This socially constructed version of the original Frayer model supports conceptual understanding by giving students an opportunity to share their thinking, listen to the ideas of others, support their ideas or refute the ideas of others using reasoning, and revise or refine their ideas as they assimilate new information. The chart can be revisited after instruction with students deciding what sticky notes to leave on the chart; which ones to remove; and as a class come to a consensus as to the best operational definition, characteristics, and examples and non-examples.

**Figure 8.7. Group Interactive Frayer Model Chart for Concept of a Rock**

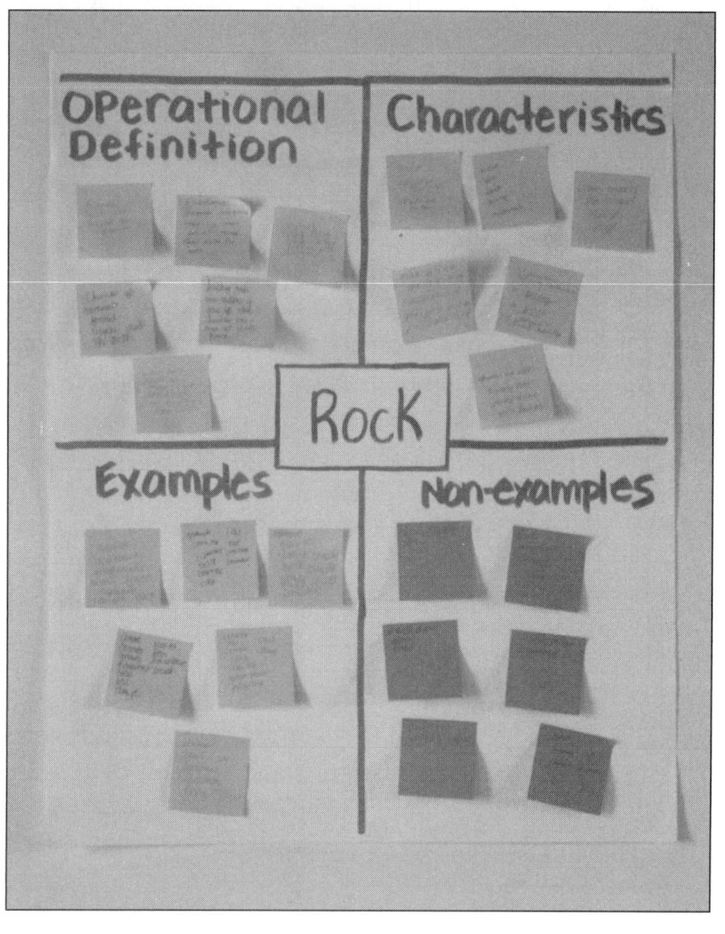


## Strategy #8: KWL

The letters *K*, *W*, and *L* stand for "What I Know," "What I Want to Know," and "What I Learned" (Ogle 1986). It was developed by Donna Ogle to help reading comprehension. Teachers have used this strategy to lead discussions and inquiry in all disciplines for many years. Technically, it should be called a graphic organizer that helps students to organize their prior knowledge, their questions about their knowledge, and what they have learned, but we include it here since it is a popular model for engaging students and using their prior knowledge to lead the lesson, their desires for further study, and a reflection on what they have learned. To begin, the teacher uses a chart with the above categories listed across the top. "What I Know" is a type of elicitation of what students bring to the classroom. (If the teacher asks, "What do you think you know about … ?" she might get more answers because it allows for a broader spectrum of experiences.) The teacher writes these in the space allocated to "Know" without comment or judgment. Often, the students and the teacher pick one "Know" comment and engage in an activity. For example, if a comment in the "Know" column says that air has weight, the teacher might compare the weights of a deflated basketball with an inflated one. The results are translated into a statement that is recorded in the "Learn" column. The students are then questioned about what they learned.

Again, with time and usage, iterations are found to improve the original model. KLEW is one such improvement on the KWL model. This was developed by a group of educators in 2006 (Hershberger, Zembal-Saul, and Starr). These innovators inserted an *E* after the

Learn column in which students stated their evidence for the *L*. The *W* was then changed to "Wonderings," which allowed the students to consider other questions based on their "Learnings" that they would like to pursue. Thus, KLEW allows for the insertion of the practice of gathering evidence and for extension of the investigation.

## Strategy #9: Our Best Thinking Until Now

Our Best Thinking Until Now can technically be considered a graphic organizer. It was developed to embellish the discussions that occurred during the use of the *Everyday Science Mysteries* series. Students are asked to list their "best thinking until now" in a conspicuous place after reading the open-ended mystery from the series. The "until now" is explained as an invitation for the students to make statements that are open to inquiry and therefore to make changes whenever they have evidence to do so. For example, in a story about pendulums, they may list some possible variables that affect the period of the pendulum, thinking that the mass of the pendulum changes the period of the pendulum (the time it takes for the pendulum to swing to and fro one time). When there is a series of "best thinkings" listed, the students are asked to change them into questions that can be tested. Thus, the statement becomes, "Does the mass of the pendulum change the period of the pendulum?" The students now have a series of questions for which they must design tests and collect data that become evidence for any subsequent claims.

## Strategy #10: Predict-Observe-Explain (POE) Sequences

A powerful way to promote conceptual change is to have students make predictions that are often based on their commonsense interpretations of everyday phenomena. Students then observe the phenomenon and often find that their observation is at odds with their prediction and interpretation of the concept underlying the phenomenon. As a result they have to modify or discard their initial ideas to try and understand the phenomenon in order to construct a scientific explanation.

POE sequences embrace many of the features of constructivist learning. They have been extensively researched and evaluated by Haysom and Bowen (2010) and found to enhance both students' and teachers' understanding of science concepts. However, we are reminded that researchers have found that students' ideas are often resistant to change and there is no guarantee that a POE sequence will change their ideas, even though it might provide the starting point for building a bridge between where students are in their thinking to where we want them to be.

Seven steps in using POE sequences are described in *Predict, Observe, Explain: Activities Enhancing Scientific Understanding* (Haysom and Bowen 2010).

- Step 1: *Orientation and Motivation.* The POE usually begins by drawing on students' past experiences or prior knowledge and initiates a question that can be tested through experimentation. The teacher engages students in an initial discussion that will provide the students with the opportunity to surface and reflect on their initial ideas.

- Step 2: *Introducing the Experiment.* The teacher introduces the experiment and links it to the students' previous discussion.
- Step 3: *Prediction: The Elicitation of Students' Ideas.* Before the students launch into the experiment, they are asked to write down what they predict will happen, along with the reasons for their prediction. This is important for two reasons: (1) Making their reasoning explicit helps the students become more aware of their own thinking, and (2) it provides the teacher with insight into where students are in their initial thinking and helps inform instructional planning.
- Step 4: *Discussing Their Predictions.* There are two stages to this process. First, the students share their predictions in a whole-class discussion, highlighting their range of predictions and the reasons for them. The teacher is careful not to pass judgment on whether their predictions are "right" or "wrong" so all students will be encouraged to share their thinking. All ideas are valued because they represent a student's best efforts to make sense of their natural world at this point. You might also mention to students that making their predictions explicit helps them learn. After students have shared their predictions, invite the class to discuss which predictions and reasons they now think are best. By providing students with an opportunity to evaluate others' reasoning and reconsider their own reasons, some may begin to change their ideas and reconstruct their thinking. You might even consider taking a straw vote at this point.
- Step 5: *Observation.* At this point the students are ready to make observations to test their predictions. You may choose to do a demonstration or have the students design and carry out their own experiments. If you choose the former, it is important to make the demonstration interactive and invite students to help you conduct the demonstration. Make sure students record their observations during the demonstration or experiment.
- Step 6: *Explanation.* After students make their observations, provide them with an opportunity to discuss their observations in pairs or in small groups. After students have had an opportunity to discuss their observations, they then construct their written or verbal explanations.
- Step 7: *Providing the Scientific Explanation.* Once students have had the opportunity to construct and share their own explanations, introduce the scientific explanation by saying, "Here is how a scientist would explain this." Provide an opportunity for the students to compare their own explanations with those of scientists, looking for similarities and differences. This provides another opportunity for them to solidify or reconstruct their ideas.

You can develop your own POE situations or use the collection published by NSTA Press. Figure 8.8 below shows an example of a POE on magnetic force.

### Figure 8.8. POE on Magnetic Force

## ***Does Magnetic Force Penetrate All Materials?***

You can stick notices and other stuff to your fridge with a magnet. Do you know what stuff sticks and what doesn't? Do you have any ideas why?

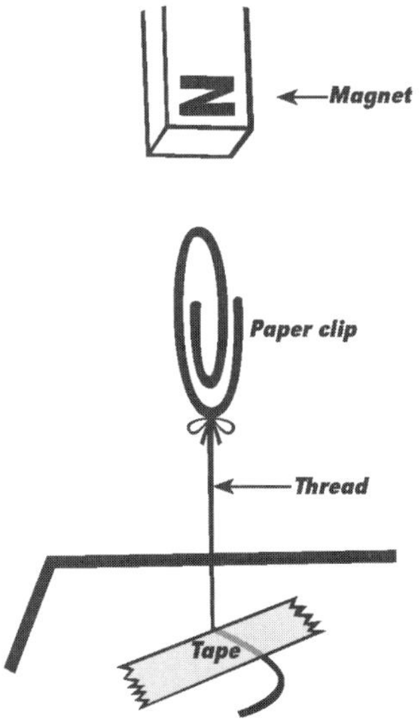

**An Experiment**
Which materials let magnetic force pass straight through them? Which don't?

- Set up a magnet so that it holds a paper clip about 1 cm from the magnet.
- Try placing different materials between the magnet and the paper clip.

**Predict**
Place a check [√] next to those materials that you think will let the magnetic force pass straight through (the ones you can stick to your fridge).

Card [ ] Aluminum [ ] Glass [ ] Iron [ ] Plastic [ ] Nickel [ ] Your choice ________ [ ]

**Observe**
Check it out!

Materials that let magnetic force pass straight through: ______________________

Materials that don't let it pass straight through: ______________________

**Explain**
What do materials that don't let the magnetic force pass straight through have in common?

______________________

______________________

***Hey!***
Do you have any ideas about why some materials don't let magnetic force pass straight through them?

______________________

______________________

______________________

*Source:* Haysom and Bowen 2010, p. 83.

## Strategy #11: Role-Playing

Some teachers at both elementary and secondary levels have found role-playing to be an excellent strategy for engaging students in a new idea. Role-playing involves the students by asking them to "become" a mitochondria, a gas molecule, an organism, or a rolling ball. This appeals to some students but not all, at first. In our experience, the older students resist a great deal more than younger students. A high school biology teacher in Connecticut introduced the term "bio drama" to her students, and after a few false starts her students really enjoyed becoming mitochondria and "feeling" the activity of energy transformation internally.

In another instance, a first-grade teacher in New Hampshire noticed that her students were stepping on any insect or spider they could. She countered by presenting each student with a mealworm (*Tenebrio molitor*) in a plastic container. She asked her students to consider it a pet and to observe it over a few weeks. Students were able to see the transformation to pupa and then to beetle. She noticed a distinct difference in the way the students treated small animals from then on. Instead of killing them, they came to her and asked for a magnifier and began to observe them. She then asked the students to "become" the mealworm and to go through the life cycle. They became eggs, emerged into larvae, ate, and then pupated and transformed into beetles in dramatic fashion. They left little doubt that they understood the life cycle of their pet and were ready to consider the concept of cycle through observing other living things.

## Strategy #12: RECAST Activities

One of the crosscutting concepts that students use to make sense of and explain phenomena is cause and effect. *A Framework for K–12 Science Education* describes cause and effect in this way: "Events have causes, sometimes simple, sometimes multifaceted. A major activity of science is investigating and explaining causal relationships and the mechanisms by which they are mediated. Such mechanisms can then be tested across given contexts and used to predict and explain events in new contexts" (NRC 2012, p. 84). The Understanding of Consequence Lab at the Harvard Graduate School of Education has studied *causal cognition*—how people understand and use causal relationships in their reasoning. Even though their research has shown that students are capable of understanding complex causality to a greater extent than what was previously believed, many of their misconceptions in science have been linked to their difficulties in understanding causality (Perkins and Grotzer 2005).

The Understandings of Consequence Project demonstrated that part of the problem in students conceptually understanding science concepts arises from differences in how students and scientists think about cause and effect. Scientific explanations often require students to make a more complex set of assumptions about how causes and effects behave than students typically do. To address student difficulties with understanding causal patterns,

Tina Grotzer and her colleagues developed the strategy of using RECAST activities. Their National Science Foundation-funded Causal Patterns in Science Project describes RECAST activities on their website (*www.causalpatterns.org/recast/defined.php*) as "activities designed to **RE**veal **CA**usal **ST**ructure or help students RECAST their explanation by drawing their attention to the underlying causal structure. By showing outcomes that don't fit with a simple causal pattern, students realize that a different causal pattern is involved."

In addition to helping students overcome commonly held misconceptions, RECAST activities deepen students' understanding of several of the crosscutting concepts such as Patterns, Cause and Effect, and Systems and System Models. Furthermore, they strengthen students' ability to evaluate arguments and distinguish between scientific causal claims and nonscientific causal claims.

For example, understanding density involves relational causality: the relationship between the mass of a unit of material and the volume of that unit of material. Neither mass nor volume alone determines density. Students need opportunities to reason about the relationship between mass and volume and observe that if the relationship between them changes, so does the density. Certain teaching practices compound the conceptual problems students have understanding density. For instance, the common practice of calculating specific densities for various materials without letting students know that density can change (such as when a substance changes from one state to another), or referring to certain objects as floaters or sinkers without reference to the liquid they are in contribute to students' misconceptions and lack of recognizing a causal relationship.

A RECAST unit on density introduces three causes that contribute to density: (1) atomic mass, (2) the strength and structure of atomic and molecular bonds, and (3) mixed density. As the module "Causal Patterns in Density" explains, "In any given instance, these causes are possible contributors to density" (President and Fellows of Harvard College 2005, p. 8). Therefore, the RECAST lessons in parts of the module focus on how to consider what it means to have multiple contributing causes. For more information about RECAST activities and example of curricula that use the RECAST approach, including *Causal Patterns in Density,* visit *www.causalpatterns.org/resources.php*.

## Strategy #13: Talk Moves

Language plays a critical role in helping students reflect on and develop their conceptual understanding. When teachers use purposeful strategies to support students' use of language in science, they guide them toward greater conceptual understanding as well as support their ability to use the language of science to explain concepts. *Ready, Set, SCIENCE!* describes the importance of learning science through talk and argument:

> *In order to process, make sense of, and learn from their ideas, observations, and experiences, students must talk about them. Talk, in general, is an important and integral part of learning, and students should have regular opportunities to talk through their*

> *ideas, collectively, in all subject areas. Talk forces students to think about and articulate their ideas. Talk can also provide an impetus for students to reflect on what they do—and do not—understand. This is why many seasoned teachers commonly ask students to describe terms, concepts, observations in their own words. (Michaels, Shouse, and Schweingruber 2008, p. 88)*

As students grapple with ideas in science, the role of the teacher is to facilitate productive science talk in ways that will move students' thinking forward and help clarify and expand on their reasoning. Even though it is widely known that talk and argument is important in learning science, observations of K–8 classrooms reveal that they are typically lacking rich opportunities for students to engage in productive science talk. The common pattern of practice is that the teacher asks a question, usually with a predictable answer, and the student is called on to respond. The teacher comments on the student's response, and then proceeds to ask another question and the pattern repeats itself, much like playing Ping-Pong: teacher to student, teacher to student, teacher to student. One of the ways to break this pattern (commonly referred to as IRE: initiate, respond, and evaluate) is to ask questions and orchestrate discussions through a set of talk strategies called talk moves. In *Ready, Set, SCIENCE!* Michaels, Shouse, and Schweingruber describe six different talk-facilitation strategies that teachers can use to encourage productive talk and argument in the classroom:

1. *Revoicing.* This talk move involves the teacher in restating a student's ideas in the form of a question. Sometimes it is difficult to understand what the student is trying to say when they struggle to put their thoughts into words. If the teacher has difficulty understanding what a student is trying to say, then the students in class who are listening are apt to have even greater difficulty. It is important to help students clearly express their ideas when they share their thinking. Therefore, this move not only helps the student clarify his or her thinking but also provides clarity for the teacher and the students in the class who are listening as well. By restating a student's idea as a question, the teacher is giving the student more time to clarify his or her ideas. It is also a strategy for making sure the student's idea is accessible to the other students who are listening and following the discussion. Examples of revoicing are: "So let me see if I got your thinking right. You're saying that …" and "I want to make sure I understand what you are thinking. Your idea is that …"
2. *Asking Students to Restate Another Student's Reasoning.* This talk move is similar to revoicing, except instead of being done by the teacher, it is done by the students. This move has the students reword or repeat what other students share during a class discussion. It is then followed up with the student whose reasoning was repeated or reworded. When this talk move is used, it gives the students in the class more think time and an opportunity to process each contribution to the science talk. It also provides a version of the explanation in student language that

may be an easier version for some students to understand, especially English language learners. It also acknowledges to the students that the teacher and the other students in the class are listening to one another. Examples of this talk move are "Can someone repeat in their own words what Matilda just said? Is that what you meant, Matilda?" and "Does anyone have another way of saying what Henry just shared? Henry, did that capture what you were trying to say?"

3. *Asking Students to Apply Their Own Reasoning to Someone Else's Reasoning.* This talk move is used to encourage students to evaluate the claims, evidence, and reasoning of their classmates. It helps students zero in on the reasoning used to support a claim. The teacher is not asking the other students whether they merely agree or disagree with someone's claim; the students have to explain *why* they agree or disagree, drawing on concepts and principles in evaluating the appropriateness and sufficiency of the evidence. This talk will help students compare their own thinking to someone else's and, in the process, it helps them be more explicit in their own reasoning as they think through their own ideas and how they are similar to or different from others'. Examples of this talk move are "Can someone tell me why they agree or disagree with Emma's reasoning?" and "Jonah, I see that you agree with Sam. How is your reasoning similar to Sam's?"
4. *Prompting Students for Further Participation.* As different ideas emerge during a science discussion, it is important to make sure that all students have opportunities to speak and are carefully listening. The teacher uses this talk move to prompt others in the class to contribute to the discussion by agreeing, disagreeing, or adding on to what was already shared. It encourages all students to evaluate the strength of each other's arguments and promotes an equitable accountable discussion. It ensures that students are always thinking throughout the discussion. Examples of this talk move are "Would someone like to add on to Tamica's explanation?" and "What do others think about the ideas that have been shared so far?"
5. *Asking Students to Explicate Their Reasoning.* This talk move encourages students to go deeper with their reasoning and be more explicit in describing the evidence, concepts, or scientific principles they use in their explanations. It helps students focus on the evidence that supports their claim and helps them build on the reasoning of others. As they are encouraged to provide more detail for their reasoning, they work harder to access and use the concepts and scientific principles they are learning about in class or of which they have prior knowledge. Examples of this talk move are "Tell us more about what you agree with Lucy"and "Can you say more about how that evidence supports your claim?"
6. *Using Wait Time.* Unlike the other talk moves described above, this talk move is actually a silent move. One of the hardest things for teachers to do is to refrain from immediately commenting on a student's response. Wait time originated from research done by Mary Budd Rowe (1974) in which she found that teachers tend to

leave no more than one second of silence before addressing an unanswered question or asking someone to answer a question. Wait time is the interval between the time a question is posed and the time either a student or the teacher responds to the question. When wait time is increased to at least three seconds (preferably five seconds), participation in a discussion increases, student responses are more detailed, and complex thinking increases. By practicing wait time, the teacher supports students' thinking and reasoning by providing more time for them to construct an explanation or evaluate the arguments of others.

These talk moves can be used with a variety of instructional materials to facilitate productive talk in which all students are held accountable for each other's learning. However, to use these moves effectively, it is important to establish the conditions for a respectful learning environment. The Inquiry Science Project (*http://inquiryproject.terc.edu*) includes excellent materials to help teachers use talk moves effectively and build a culture of productive science talk in the classroom (click on the "Talk Science" box). Another excellent resource for supporting discourse in the science classroom is the Tools for Ambitious Science Teaching website (*http://tools4teachingscience.org*).

## Strategy #14: Thinking About Thinking: Supporting Metacognition

Metacognition is the process of being aware of one's thinking. Several researchers have found that students' metacognitive abilities may be a critical factor in achieving conceptual change (Minstrell 1989; Beeth and Hewson 1999; Gelman and Lucariello 2002). While teachers guide students through a process of conceptual change, students also have an important role to play in the process by becoming self-regulated learners. Self-regulated learners are able to set goals, find strategies that help them achieve those goals, and monitor their progress toward learning scientific ideas (Schraw, Crippen, and Hartley 2006). Metacognition involves reflection as well as self-monitoring.

## Author Vignette

As a middle school teacher, I was interested in helping my students become more aware of their thinking and used various strategies to help them become more metacognitive. For example, I frequently asked students to "think out loud," go "MTV" (make your thinking visible), or "tell me why you think that." I often used formative assessment scenarios that involved characters expressing their different ideas, such as "Food for Corn," that helped students think about their own ideas and how they were similar to or different from the views expressed by the characters (Keeley 2011).

### Food for Corn

Eight farmers were talking about their cornfields. They each had different ideas about the food their corn needed to grow. This is what they said:

**Mrs. Farrin:** "My corn plants use sunlight as their food."

**Mrs. Tobias:** "My corn plants use food they get from the soil."

**Mr. Cullenberg:** "My corn plants use sugar as their food."

**Mr. King:** "My corn plants use food from the fertilizer I give them."

**Mrs. Joslyn:** "My corn plants use carbon dioxide and water as their food."

**Mr. Cody:** "My corn plants use food from the chlorophyll in their leaves."

**Mr. Trask:** "My corn plants use food from the ears of corn they produce."

**Mrs. Ahlholm:** "My corn plants don't use food; instead, they make food for animals to eat."

Which farmer do you agree with the most? _______________ Explain why you agree with that farmer.

These eventually lead to one of my signature formats for the *Uncovering Student Ideas in Science* series, the "friendly talk probe." The use of this format helps students see the importance of thinking through

> our ideas and making them visible to ones' self and to others. Not only did the use of metacognitive teaching strategies help students surface and think through their own ideas, they also provided information to me about next steps to further develop students' ideas and help students monitor their own progress toward achieving scientific understanding.
>
> —Page Keeley

There are various techniques teachers can use to help students become more metacognitive. Here are some general suggestions:

- Create a learning environment that safely encourages students to share their thinking through interactive discussions with peers (not just back and forth with the teacher) and engage in argumentation to support their ideas.
- Elicit initial predictions then provide follow-up opportunities through investigation or demonstration to test students' predictions. Then use discussion to promote thinking that helps students reconcile the difference between their prediction and observation or even if their prediction is correct, discussion helps solidify their thinking.
- Use conceptual checks throughout a sequence of instruction, asking students what they are thinking now and why.
- Provide opportunities for reflection, encouraging students to identify and explain what led them to change their thinking.

## Strategy #15: Thought Experiments

A thought experiment is a hypothetical scenario that helps students think through their ideas and access concepts to predict and explain the outcome of an imaginary situation. Thought experiments require students to think through and apply ideas since they usually cannot be empirically tested. Thought experiments have been used by scientists as "proxy experiments." Famous thought experiments include Schrödinger's cat, Maxwell's demon, and Galileo's dropping of heavy and light balls.

Thought experiments, by their very nature of being imaginary, engage students in making predictions supported by explanations. As they alternate between the imaginary situation and their own mental models and prior investigations, they think through how a phenomenon or system works. The mental imagery involved creates a high level of cognitive engagement (Gilbert and Reiner 2000). Thought experiments often encourage students to draw their conceptual models, physically model similar situations, use mathematical thinking, and engage in rich discourse with their peers to explain their ideas. While thought experiments have been used mostly in physics education, they can be designed and used with any grade level. For example, the formative assessment probe "Falling Through the Earth" (Figure 8.9) is used to explore middle school students' ideas about gravity (Keeley and Sneider 2012). Students can even develop and share their own thought experiments to explore ideas about concepts.

Thought experiments promote conceptual change by supporting an existing explanation, challenging an explanation, or leading to the development of an explanation. Research has found that the construction and use of thought experiments have been found to help students solve problems, increase collaboration among students, and communicate their thinking to others (Reiner 1998), allowing students to act in much the same way as scientists when encountering perplexing situations.

### Figure 8.9. "Falling Through the Earth" Probe

**Falling Through the Earth**

A teacher asked her students to imagine that it was possible to drill a hole all the way through the Earth from the North Pole to the South Pole. The hole is lined with super-strong steel so that it does not collapse or melt. There is air inside the hole. She asked the students to discuss what would happen to a rock that is dropped into the hole. Here is what they said:

**Alana:** "It would fall into the hole and would just keep going until it hit something."

**Nate:** "It would just fall straight down and come out the other side."

**Tess:** "I bet it would come out the bottom of the Earth and just keep falling forever into space."

**Tim:** "It will go to the center of the hole and stop."

**Jean:** "It will pass through the center, slow down, and fall back toward the center again."

**Frank:** "It's probably just going to stick to the side somewhere."

Whom do you agree with the most? ________________ Explain why you agree.

*Source:* Keeley and Sneider 2012.

## Author Vignette

Many years ago when I taught middle school, I recall asking my eighth graders to come up with some interesting ways to further explore and develop their understanding of the concept of density as a proportional relationship between mass and volume. One student decided we could calculate the density of an average human to show why humans can float in water (and yes, we had a willing volunteer who we weighed and then submersed in a full barrel of water placed in a kiddie pool to capture and measure the volume of water displaced in order to find the volume of his body!). The floating/sinking idea triggered a thought experiment from another student who wondered what would happen to the planet Saturn if the planet could be dropped into a huge tank of water. This led students to "test" the hypothetical situation by researching the mass and volume of Saturn to calculate its density and compare it to the density of water. This further led to other hypothetical density thought experiments that captured their interest and confronted a common misconception that bigger doesn't always mean denser.

—Page Keeley

## Selecting Your Own Instructional Strategies

The examples in this chapter are just a few of the many instructional strategies that support conceptual understanding in science. You may have other strategies that you have used or perhaps heard about. In selecting an instructional strategy that moves beyond treating students like empty vessels into which we must pour all kinds of information or engaging students in hands-on activities that focus more on the materials and the "fun" of science and do little to move students' learning forward, select strategies that focus more on where students are in their thinking, and engage them in continually revisiting, revising, and refining their thinking. Ask yourself these questions as you decide whether an instructional strategy can help your students develop conceptual understanding that is more compatible with scientific ideas:

1. Will this strategy help me understand where my students are in their initial thinking?

2. Will this strategy provide a motivating experience and create a desire to learn?
3. Will this strategy make it safe for students to surface and share their thinking?
4. Will this strategy encourage students to ask questions related to a concept and encourage exploration?
5. Will this strategy provide an opportunity for students to clarify their thinking and evaluate the thinking of others?
6. Will this strategy facilitate an exchange of different views and ways of reasoning among students in the class?
7. Will this strategy provide opportunities for students to test the validity of their claims by seeking, collecting, and interpreting evidence?
8. Will this strategy provide an opportunity for students to do more of the talking than the teacher?
9. Is this strategy focused less on the materials students are using and more on the sense making?
10. Will this strategy help students work through their ideas and build a bridge between where they are initially in their thinking and where they need to be scientifically?

## Questions for Personal Reflection or Group Discussion

1. Which of these strategies have you used in your classroom? How did your use of the strategy support conceptual understanding?
2. Pick one strategy from the 15 examples provided that you would like to use in your classroom. Why did you select this strategy? How will it help you meet your instructional goals?
3. What is an example of an instructional strategy you have found helpful? How does it measure up against the list of characteristics given in this chapter?
4. Argumentation is one of the strategies listed. The practice of argumentation is also embedded within several strategies. For some children, argumentation is often about "winning." How would you go about setting norms for "arguing" so that the process achieves the effect of learning concepts for understanding and promotes civility?
5. Go to the Inquiry Science Project (*http://inquiryproject.terc.edu*) and click on the "Talk Science" link. Find some resources to enhance talk in the classroom.
6. The strategies listed involve a lot of interaction among students. Sometimes teachers are concerned about having a "noisy classroom." How would you explain

why your classroom was noisy if an administrator or visitor came to observe your classroom?

7. Eleanor Duckworth states that often children's learning is hampered by the fact the "the problem is that they don't see the problem." What do you think she means by this and how can you use instructional strategies to help them "see the problem"?
8. Take a look at the characteristics of instructional strategies that support conceptual learning. How do the instructional strategies you currently use measure up to these characteristics?
9. Choose one "golden line" from this chapter (a sentence that really speaks to or resonates with you). Write this on a sentence strip and share it with others. Explain why you chose it.
10. What was the biggest "takeaway" from this chapter for you? What will you do or think about differently as a result?

## Extending Your Learning With NSTA Resources

1. Many of these strategies can be combined with laboratory experiences that support learning. Read and discuss the article Eisenkraft, A. 2013. Closing the gap: Laboratory experiences, and not just textbooks, are the best way to provide equal learning experiences for all. *The Science Teacher* 80 (4): 43–45.
2. The NSTA journals include many articles on using analogies in the science classroom. Search the online NSTA member journal archives to find an article to read and discuss on using analogies to teach science.
3. Watch a recorded NSTA webinar on claims, evidence, and reasoning and the role of argument in science writing at *http://learningcenter.nsta.org/products/symposia_seminars/NSTA/webseminar18.aspx*
4. Read and discuss an article on how KWL is used to guide elementary students through their learning experiences: Crowther, D., and J. Cannon. 2004. Strategy makeover: From Know-Want-Learned to Think-How-Conclude. *Science and Children* 42 (1): 42–44.
5. Try out the POE examples in Haysom, J., and M. Bowen. 2010. *Predict, observe, explain: Activities enhancing scientific understanding*. Arlington, VA: NSTA Press.
6. Read and discuss Chapter 17: Thinking About Thinking in Science Class, in Hanuscin, D., and M. Rodgers. 2013. *Perspectives: Research and tips to support science education, K–6*. Arlington, VA: NSTA Press.
7. Several of the strategies explained in this chapter, such as Talk Moves, Group Interactive Frayer Model, argumentation, modeling, and more are described as

classroom vignettes in Keeley, P. 2014. *What are they thinking? Promoting elementary learning through formative assessment.* Arlington, VA: NSTA Press.

8. Search the NSTA Learning Center for information and resources on instructional strategies that support conceptual understanding (*http://nsta.learningcenter.org*).

# Chapter 9

## How Does Linking Assessment, Instruction, and Learning Support Conceptual Understanding?

Picture a second-grade classroom where students are having a science discussion about matter. The teacher uses the "Is It Matter?" formative assessment probe (Figure 9.1) to elicit students' ideas about matter (Keeley 2013a). The students look at the pictures on the list and decide which of the things are matter. They share their thinking about the characteristics that determine whether something is considered matter.

**Figure 9.1. "Is It Matter?" Probe**

*Source:* Keeley 2013b.

Some students claim that the ant is not matter because it is an insect and it is living. They all agree the rock is matter. They support the idea that rock is matter with reasons such as "you can see it," "you can hold it," and "it is hard." Some students disagree that matter has to be hard. They claim that the sponge and the water are matter because you can see them and you can hold them, but they are not hard. The teacher asks "What about ice cream? Is ice cream matter?" Some students argue it is matter using a rule they established that you could see it and hold it, even though it drips through your fingers as it melts. If you could feel it, then it is matter. "What do you think about air in the balloon? Is air matter?" the teacher asks. The class gives a resounding "no." One student says, "It can't be matter because you can't see air and you can't hold it." However, one student challenges that idea by adding that sometimes you could feel air, "like when air gushes out of a balloon." The teacher asks, "What about the house? Is the house matter?" The students decide you could see it, you couldn't hold it because it was too big, but it is still matter because you could touch it and feel it. The teacher asks if size makes a difference as to whether something is or is not matter. The students ponder this and decide it depends on how small something is. If it gets so small that you can barely see it, then it probably is not matter. Using this rule, some decide bits of dust are not matter, while others think it is matter when it is a big clump of dust.

The entire time the teacher is facilitating the science talk, she is listening carefully to the students' ideas and the reasoning they use to support their concept of matter. She does not correct them, but instead uses their ideas to inform the instructional experiences she will provide next to help the students come up with a common rule that would determine whether something is considered matter, building a foundation that will help them understand core ideas related to the structure and the properties of matter.

**Figure 9.2. "Needs of Seeds" Probe**

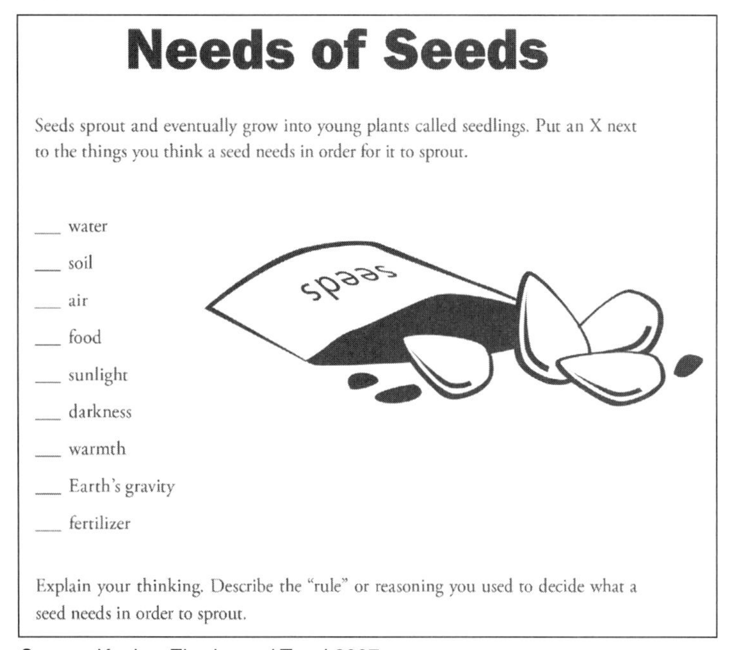

**Needs of Seeds**

Seeds sprout and eventually grow into young plants called seedlings. Put an X next to the things you think a seed needs in order for it to sprout.

___ water
___ soil
___ air
___ food
___ sunlight
___ darkness
___ warmth
___ Earth's gravity
___ fertilizer



Explain your thinking. Describe the "rule" or reasoning you used to decide what a seed needs in order to sprout.

*Source:* Keeley, Eberle, and Tugel 2007.

In a fourth-grade classroom, students are using the card sort strategy with the formative assessment probe "Needs of Seeds" (Figure 9.2). The answer choices are printed on cards and students sort the cards into three columns: things seeds need to sprout, things seeds do not need to sprout, and things we do not agree on. As the students work in small groups to sort the cards, they engage in argumentation to support their ideas. The teacher circulates throughout the class carefully listening as students share their thinking with each other. The teacher observes the placement of their cards and with a quick glance, can note understandings and misunderstandings about what seeds need to grow.

The teacher notices that many of the groups think all seeds require darkness in order to sprout. As she listens to the students, they explain how seeds need to be covered by soil, and be away from sunlight, in order to sprout and grow. When she probes the students further, they draw on their own prior experiences planting seeds in a garden or in a pot and covering them with soil. They explain that the soil keeps the light out. They also share their belief that seeds could not sprout without soil because they need something to put their roots in. The teacher uses the information obtained from listening to students discuss their ideas as they sort each card to design learning experiences that will challenge and build off of the students' preconceptions. In particular, she notes that her students will need to investigate germinating seeds to determine several factors, including whether seeds must be in darkness to sprout. She has the students design their own experiments to test their ideas about what seeds need in order to sprout. She thinks that if students have the opportunity to make their own predictions, test them by making observations, and then see whether their observations matched their predictions, they might give up some of the naive ideas they revealed during the card sort activity.

As another option, imagine a class reading "Springtime in the Greenhouse," from Konicek-Moran's book *Everyday Life Science Mysteries* (2013b). In the story, the mother is setting seed trays for her business; her sons ask if she needs fertilizer for the seeds to germinate, and they ask if the seeds need to be placed beneath the surface of the soil in order to germinate. This sets up a need for trying various seeds in a multitude of environments to answer the questions, another form of assessment. This is just another way of helping students to use actual living things to answer questions through scientific practices. It is also a way of assessing student's prior knowledge about what seeds need to germinate.

Picture a middle school classroom where the teacher uses the formative assessment probe "Earth or Moon Shadow?" (Figure 9.3) to elicit students' initial ideas about why we see different Moon phases (Keeley and Sneider 2012). After students select the person they most agree with and write an explanation for their answer choice, the teacher examines their responses in order to inform the next day's lesson on Moon phases. She notices that many of the students chose Enrique, with a common explanation being that the shadow of the Earth is cast on the Moon and at different times of the month the amount of the shadow coming from the Earth changes. She is aware that this is a common misconception that many students, as well as adults, have. She develops the learning intention for the next day's lesson so the students know the purpose of the lesson is to understand what causes the phases of the Moon. She plans the lesson so students will use a model to observe that it is not the shadow of the Earth but rather the Moon's own shadow that results in seeing the different Moon phases from Earth. The model will help them understand that

**Figure 9.3. "Earth or Moon Shadow?" Probe**

**Earth or Moon Shadow?**

Two friends were looking at the Moon. Part of the Moon was visible to them. They wondered why they could only see part of the Moon. This is what they said:

**Sally:** "I think the part we can't see is the Moon's own shadow."

**Enrique:** "I think the Moon has moved into the Earth's shadow."

Circle the friend you agree with the most: Sally Enrique

Explain why you agree. ____________________

*Source:* Keeley and Sneider 2012.

the part of the Moon we see is due to the positional relationship of the Moon in relation to the Earth and the Sun. The teacher develops three success criteria that will help students self-monitor their own learning as they develop and use a model to explain what causes Moon phases.

As an alternative, imagine a group of students reading "Moon Tricks" from Konicek-Moran's book *Everyday Earth and Space Science Mysteries* (2013a). In this story, a child moves into a new house and notices at bedtime that the full Moon is shining in his window. The next night there is no moonlight and this continues for a week when in the middle of the night he is awakened from his sleep to see a Moon in his window that is minus the right half. He talks with his family the next morning, each member having a different explanation for what the child has seen. Each explanation is based on common alternative conceptions. In a class discussion, students agree or disagree with certain members of the family. These distractors help the teacher see how her class views the Sun-Moon-Earth relationship, and usually leads to a lively discussion. The teacher suggests that the class keep a Moon journal of shape, size, direction, and time, each night and then try to come up with an explanation of the Moon's behavior. After the journal data are discussed, the teacher helps the students come up with a model of the Sun, Moon, and Earth that will explain the observations.

**Figure 9.4. "What's in the Bubbles?" Probe**

**What's in the Bubbles?**

Hannah is boiling water in a glass tea kettle. She notices bubbles forming on the bottom of the kettle that rise to the top and wonders what is in the bubbles. She asks her family what they think, and this is what they say:

Dad: "They are bubbles of heat."

Calvin: "The bubbles are filled with air."

Grandma: "The bubbles are an invisible form of water."

Mom: "The bubbles are empty—there is nothing inside them."

Lucy: "The bubbles contain oxygen and hydrogen that separated from the water."

Which person do you most agree with and why? Explain your thinking.

*Source:* Keeley, Eberle, and Tugel 2007.

Picture the first day of a high school physical science unit on chemical and physical changes. Students are asked to complete the "What's in the Bubbles?" formative assessment probe (Figure 9.4). Although this seems like a simple basic question that high school students would know after learning about states of matter and the difference between a physical and a chemical change in middle school, the teacher knows that many of his students come to their high school science class with alternative ideas that were never challenged in previous grades. After the students complete their own individual response to the probe and turn it in to the teacher, the teacher then has the students talk in pairs to share their answer choices and their explanations. As the students discuss their ideas, the teacher quickly scans through the student papers to get a sense of where the class is in their thinking. After students share their thinking in pairs, the teacher opens up the discussion to the whole class. The teacher is surprised to find out that many of the students think the bubbles are filled with air or are empty and contain nothing. A few students even explained that the outside of the bubble itself was made up of water but it was empty inside. During the whole-class discussion, some students supported their claim that the bubbles are actually water vapor, the gas form of water. As students considered the evidence and reasoning for this claim, this claim became more plausible to some of the students who had selected other answer choices. Rather than give the students the answer, the teacher explains that they will discover the answer for themselves over the next few days. In the meantime, he lets the students keep thinking.

The teacher carefully plans his instruction so students will encounter their ideas through the lab activities and information from text and classroom discussions. After students complete a set of lessons on physical change, including change in state, the teacher returns their papers and gives them an opportunity to change their answer choice and support it with a revised explanation. The lesson ends with a reflective conversation to discuss how their ideas had changed and the evidence that led them to change or expand their thinking.

What do all four of these grade-level examples have in common? Each of these teachers is using formative assessment to find out what their students really think about a concept or a phenomenon prior to launching into instruction. They are using formative assessment throughout their sequence of instruction to keep their finger on the pulse of students' conceptual learning by monitoring the extent to which their students' ideas are changing to be more aligned to scientific ways of thinking. They are providing opportunities for their students to use metacognitive strategies to take responsibility for their own learning. Reflection is used to look back on initial ideas and recognize how and why one's own thinking has changed.

These teachers are linking assessment, instruction, and learning by using formative assessment to inform their instructional decisions while at the same time encouraging students to think conceptually and consider each other's ideas and ways of reasoning. Each example shows the inextricable link between assessment, instruction, and learning. The teachers are using assessment *for* learning rather than assessment *of* learning. The preposition makes a difference. This process that seamlessly links assessment, teaching, and learning is called *formative assessment*.

The primary purpose of formative assessment is to inform instruction while simultaneously promoting learning. There is a strong and substantive body of research that supports formative assessment as one of the most important and effective educational practices to support learning. John Hattie's groundbreaking metasynthesis of more than 15 years of research involving millions of students and representing the largest collection of evidence-based research into what actually works in schools to improve learning ranked formative assessment (he refers to it as formative evaluation) as #4 out of 150 studied educational influences, in terms of effect size (Hattie 2012). Linking instruction and assessment as if they were two sides of the same coin is one of the most powerful ways to support conceptual teaching and learning. Unlike summative assessment, which is used post instruction to measure and document what students have learned, formative assessment is used throughout an instructional cycle to determine where students are in their thinking and understanding at any point in time so that the teacher can adjust or modify instruction based on the learning needs of the student. Formative assessment is a broad area. For the purposes of this book, we will focus on formative assessment as it is used to uncover and understand students' thinking.

## Linking Assessment, Instruction, and Learning With an Instructional Model

In Chapter 7 we discussed how the use of instructional models supports conceptual understanding in science. What would it look like if assessment were embedded within the two most commonly used instructional models, the 5E Instructional Model and the Conceptual Change Model (CCM)?

### *Engage (5E) and Commit to an Outcome, Expose Beliefs (CCM)*

During these stages of instruction, the teacher develops curiosity, and stimulates student interest in the content of the lesson. One widely accepted role of any teacher who teaches for conceptual understanding is that of student motivator. Motivation to learn begins with questions, puzzles, stories, and problems that are interesting to students. This stage also focuses students' attention on the content to be learned. Osborne and Freyberg (1985) describe that the problem in teaching science is often not one of getting students' attention, but helping them attend to the "right" things. Science lessons often involve interesting phenomenon and events, which may be viewed in many different ways by students. The focus that the teacher intends is not always the same as the one adopted by students. During this stage the teacher shares the learning intention so that students know what the lesson will be focused on. At the same time the teacher uses strategies to reveal students' interest in the topic of the lesson, and their readiness to build on prior learning by gathering information about prerequisite learning goals and students' prior experiences that prepare them to be motivated to learn new ideas.

Elicitation of students' initial ideas is an integral part of these stages of instruction. Uncovering the initial ideas students have developed through their prior experiences, intuition, and interpretations of familiar phenomena provides a starting point from which the teacher can design instruction that will build a bridge between where students are in their initial thinking to where they need to be in their scientific understanding. At this stage, teachers become aware of the ideas that students bring to their learning that may affect their understanding of science concepts. At the same time students activate their own thinking, surface their ideas, share their thinking with others, and consider the initial ideas of others and compare them to their own. At this stage, students are making their thinking visible to themselves, their peers, and the teacher. The *Uncovering Student Ideas in Science* and the *Everyday Science Mysteries* series are excellent resources teachers can use to elicit students' initial ideas as a starting point for developing conceptual understanding.

### *Explore (5E) and Confront Beliefs (CCM)*

These stages involve direct experience with phenomena or objects, making predictions and testing them, accessing information from text or other resources, and discovering alternative ways of thinking through discussion and argument. Teachers observe students and listen carefully during this stage to determine the kinds of understandings and questions students have before developing more structured opportunities to formalize conceptual learning. Formative assessment strategies are used to reveal how students are responding to the instructional activities, how they are considering the ideas of others, and whether their initial ideas are changing or being restructured during their exploratory experiences.

### *Explain (5E) and Accommodate the Concept (CCM)*

During these stages the teacher introduces and formally develops the concepts, skills, and scientific practices students need to understand and use in order to move their learning forward toward a scientific understanding. Students engage in sense making, clarification, and solidification of ideas and processes. Discrepancies between their existing ideas and their observations or information obtained from their exploration of the content are resolved through further probing and discussion. Assessment strategies are used to determine the extent to which students are grasping a concept, speaking or writing with appropriate terminology, or using a skill or scientific practice. The information from formative assessment informs next steps for instruction by identifying the need for additional learning experiences to build more solid understandings, indicating readiness to link formal terminology to a concept, or signaling that students are ready to transfer their ideas to a new context. Feedback further enhances the opportunity to develop and refine conceptual understanding.

### *Elaborate (5E) and Extend the Concept and Go Beyond (CCM)*

During this stage, students use their newly formed or modified concepts in a new situation or context. Sometimes student understanding is limited to the context or examples in which they learned the concept. Assessment during this stage reveals whether students are able to transfer what they learned to different examples and situations. "Students' abilities to transfer what they have learned to new situations provides an important index of adaptive, flexible learning; seeing how well they do this can help educators evaluate and improve their instruction" (Bransford, Brown, and Cocking 2000, p. 235). These stages also provide an opportunity for students to raise and pursue additional questions related to the concept. Student questions provide the teacher with additional insight into their thinking.

### *Evaluate (5E)*

This stage provides an opportunity for students to use self-assessment and reflect on their learning. It can happen at the end of a sequence of instruction as students look back and think about what they have learned. It is also used throughout a cycle of instruction to help students develop important metacognitive skills that enable them to monitor their own thinking and learning. Students learn to *think about learning* as well as *think about thinking*. Self-assessment helps students think about whether the *concept makes sense*. Alternatively, reflection helps them think about how they *make sense of the concept*. Information is fed back to the teacher to inform how students' ideas have changed or deepened over the course of instruction, from a single lesson to a full unit, as well as the extent to which students are aware of what they have learned.

## Formative Assessment Strategies That Support Conceptual Understanding in Science

Formative assessment is a process that informs instruction and supports learning, with instructional decisions made by the teacher or learning decisions made by the student being at the heart of the process. Dylan Wiliam describes the big idea of formative assessment as "evidence about learning is used to adjust instruction to better meet students' needs—in other words, teaching is *adaptive* to the learner's needs" (Wiliam 2011, p. 46). Formative assessment can be broken down into five key strategies (Leahy, Lyon, Thompson, and Wiliam 2005):

1. Making the learning target clear and explicit to students
2. Designing and facilitating productive classroom discussions, activities, and tasks that elicit evidence of where students are in their learning
3. Providing feedback that moves learning forward
4. Activating students as instructional resources for one another
5. Activating students as the owners of their own learning

Formative assessment is so inextricably linked to teaching and learning that it is often difficult to determine whether the teacher is using an instructional or assessment strategy. Students interact with formative assessment in a variety of ways—through writing, drawing, speaking, listening, and carrying out investigations. Unlike formal summative assessments, such as quizzes and tests, formative assessment strategies are seamlessly embedded in instruction. They are used prior to instruction, during lessons, and even at the end of a lesson or series of lessons for reflection. By linking instruction and assessment for the purpose of informing teaching and promoting learning, teachers can

- uncover students' initial ideas about core disciplinary ideas and concepts they will encounter in their science lessons;
- stimulate intellectual curiosity and a desire to learn;

- analyze their own teaching by examining how students are progressing toward conceptual understanding of scientific ideas;
- launch into student inquiry and investigation;
- engage students in productive science talk;
- determine whether students need to experience or use ideas and concepts in different contexts;
- signal when students are ready to learn more advanced concepts, or when they need to go back and revisit fundamental concepts;
- help students recognize that their ideas are valued and taken into account by the teacher to design instruction that meets their learning needs;
- provide useful feedback that can be acted upon for improvement;
- support the development and use of academic language in science;
- encourage students to take charge of and monitor their own learning;
- identify students who can support the learning of their peers who need help in moving toward a learning target;
- focus students on the purpose of an activity, lab, reading, or any other part of the lesson; and
- provide an opportunity for students to reflect on their own growth in understanding.

There are many different types of formative assessment tools and strategies that can be used for the formative purposes described above. Both of the authors of this book have developed and published formative assessment tools and strategies that are widely used by teachers and have been included as examples throughout this book. These include the formative assessment probes from the *Uncovering Student Ideas* series and the stories from the *Everyday Science Mysteries* series. In addition, the two volumes of *Science Formative Assessment* (Keeley 2008, 2015) collectively provide 125 different formative assessment classroom techniques in science (FACTs). These formative assessment resources strongly support findings from research including:

1. "If [the students'] initial understanding is not engaged, they may fail to grasp new concepts and information presented in the classroom, or they may learn them for purposes of a test but revert to their preconceptions" (Bransford, Brown, and Cocking 2000, p. 14). The formative assessment probes, mystery stories, and collection of FACTs are used to elicit the ideas students bring to their learning before engaging in an investigation or other learning experience. By knowing in advance the preconceived ideas students have, the teacher can design targeted instruction and monitor students' learning during the lesson or sequence of lessons.

2. "A metacognitive approach to instruction can help students learn to take control of their own learning by defining learning goals in monitoring their progress in achieving them"(Bransford, Brown, and Cocking 2000, p. 18). Each of the probes and mystery stories are designed to target disciplinary core ideas and scientific practices. Students are encouraged to make their thinking visible and explicit as they interact with these formative assessment resources. Through the use of metacognitive FACTs, students oversee and steer their own learning toward a learning target.
3. "Students have to be active in reconstructing their ideas and then to merely add to those ideas an overlay of new ideas leads to poor understanding, if not to confusion" (Black and Harrison 2002, p. 4). Learning must be done by the student; it cannot be done by the teacher. Filling students' heads with new information or engaging in hands-on activities that lack a connection to students' existing ideas does little to promote conceptual understanding. Formative assessment probes, stories, and FACTs actively engage students in the process of continuously examining their thinking and helping them realize when their ideas need to be modified or discarded in favor of an alternative idea.
4. "Creating a classroom culture where students feel they can reveal their understandings and be helped to firmer understanding is an essential ingredient to making formative assessment function in the classroom" (Black and Harrison 2002, p. 5). Formative assessment probes, mystery stories, and FACTs encourage peer discussion, which is important to creating a supportive learning environment. The opportunity to discuss and argue their ideas in pairs or small groups helps students articulate their ideas before sharing them with the whole class. It provides an opportunity for the student to work out what they know, partially know and do not know about a specific idea or concept. Being able to safely discuss ideas with peers brings the students' own ideas and ways of thinking to the surface.

## Author Vignette

Teachers often ask me how I got interested in formative assessment and what led me to develop the formative assessment probes in the *Uncovering Student Ideas* series. They are surprised to find out that my formative assessment work started when I read an article in *The Phi Beta Kappan* back in the early 1990s. I was a middle school science teacher at the time and was considered to be an exemplary inquiry-based science teacher. My classroom was a model of inquiry-based teaching at that time. However, when I read the article, I realized what was missing in my teaching. My students had opportunities to do a lot of interesting hands-on science investigations and my classroom was always abuzz with activity, but these activities were planned around my ideas of what I thought my students needed. After reading the article, *Teaching for Conceptual Change: Confronting Students With Their Experience* (Watson and Konicek 1991), I realized I needed to start my lessons from where my students were, not from what I thought would be a good activity. I decided to try the same approach as the teacher in the case study described in the article. I started a lesson by asking my students what would happen to the temperature reading on the thermometer if it was put inside a mitten for a couple minutes. My students were convinced that the temperature reading on the thermometer would go way up since mittens are used to warm up our hands. I did not correct them but rather let them test their prediction. When they saw there was no significant difference, they were just like the students in the article. They insisted the thermometer was not left inside the mitten long enough. They would not give up their strongly held misconception and believed that the temperature was not increasing because somehow cold air was getting into the mitten.

Finally, I took them outside on a cold winter day, and had them put thermometers inside their mittens and slip their hands inside the mittens. After a few minutes they took the thermometers out, and notice that this time the reading on the thermometer increased significantly. At that point my students realized they needed to discard their old idea that the mittens warmed up their hands. After some discussion, they came up with the idea that mittens trapped their body heat and slowed down the movement of heat from their body to the air. They essentially came up with their own concept of an insulator to slow down the transfer of heat. They were so excited to discover this for themselves and experience the process of conceptual change. Thus, my first probe, "The Mitten Problem," was created in 1992 and later published in my first book (Keeley, Eberle, and Farrin 2005). From that point on, I always started a lesson with an interesting question that would uncover what my students were thinking and then designed instruction that would build a bridge between their initial ideas and the science concepts and principles they needed to learn and be able to use.

My students always looked forward to these questions and enjoyed the opportunity to surface their ideas and work through them. To this day, over 20 years later, I still run into former students who remember and fondly recall the lesson "when we went outside with thermometers inside our mittens." Today over 240 formative assessment probes have been published in the *Uncovering Student Ideas in Science* series as a result of reading that article and changing my instructional practice. It was a decade later that I met the author of the article, Dick Konicek-Moran, and have been immensely privileged to collaborate with him ever since.

—Page Keeley

## Ten Suggestions for Using Formative Assessment

Now that you may be clear about the purpose of formative assessment, the research that supports it, some tools and resources for formative assessment in science, and its link to an instructional cycle, you might be wondering about some of the ways you can use formative assessment to support conceptual understanding in science. The following are some suggestions for getting started:

1. *Be clear about your purpose.* Before you use a formative assessment probe, mystery story, or other formative assessment classroom technique (FACT), be clear about the purpose. Why do want to use it? What is the core disciplinary idea or scientific practice that the formative assessment strategy will address? What will you

learn about your students' thinking if you use this strategy? How will it support student learning?

2. *Think like a diagnostician.* Students' preconceptions have a powerful influence on their learning. Therefore you need to continuously use formative assessment strategies to uncover students' ideas in such a way that it becomes second nature to you (Osborne and Freyberg 1985). Take advantage of every opportunity to explore students' ideas in-depth and analyze their thinking.

3. *Change from inquiry to inquiry for conceptual change.* Inquiry has been the centerpiece of science teaching and learning for the last two decades. However, inquiry is much more powerful in producing conceptual understanding if it begins with eliciting students' initial ideas and strategies are used to continuously monitor student thinking throughout the inquiry process.

4. *Create a classroom culture of ideas, not just right answers.* Encourage students to share their ideas, regardless of whether they are right or wrong. Most students have gone through school in a culture where they are expected to give the right answer. Thus some students hesitate to share their ideas when they think they might be wrong. Refrain from passing judgment on students' initial ideas. Provide time for them to work through their ideas and construct new understandings. Getting all ideas out on the table first maybe frustrating and seem to take longer, but in the long run, it will help your students develop confidence in sharing their explanations and result in deeper, enduring understandings.

5. *Develop a discourse community.* One of the key features of formative assessment is the way it reveals students' ideas and promotes learning through discussion and argumentation. Formative assessment does not have to be a writing activity. Sometimes it is more powerful in revealing students' ideas and promoting learning when used in a talk format. When students are talking about their science ideas, they are using the language of science as well as language that has meaning to them. "Talking the talk is an important part of learning" (Black and Harrison 2002, p. 4).

6. *Encourage careful listening.* Students need to learn to listen carefully to others' ideas and evaluate the evidence and reasoning that may lead them to change their own ideas. Teachers also need to develop the skills of careful and purposeful listening.

7. *Use a variety of grouping configurations.* Social interaction plays a powerful role in learning. While it is important to give students an opportunity to individually think through and record their ideas, it is also important for them to share their thinking with others. Students solidify their thinking and often revise or reconstruct their ideas when they have the opportunity to discuss them in pairs, small groups, and whole-class discussions. Having different learning partners when formative assessment strategies are used provides opportunities for students to

listen to and evaluate different perspectives. For this reason it is important to make sure that students are not always interacting with the same partner or in the same small group.

8. *Create a safe environment.* Formative assessment strategies often encourage students to take risks by putting forth their ideas in front of their peers. Classrooms where students feel part of an intellectual learning community that takes ownership of students' ideas, without judgment, are places where students and teachers can thrive.
9. *Don't grade formative assessments!* Formative assessment is used to reveal information about students' thinking. When formative assessment tasks are graded, students will often give the answer they think the teacher expects, rather than what they really think. Also, it is not fair to grade students on material they have not had sufficient opportunity to learn yet. Explain to students what the difference is between formative assessment and assessments that are used for grades. When students realize that you are using formative assessment to inform your teaching so that they can learn, and that they are not going to be graded on their answers, they are more apt to reveal the information you need to make formative decisions.
10. *Encourage continuous reflection.* Understanding is an evolving process. Encourage students to reflect back on their initial ideas and think about how and why their ideas have changed. It takes time for students to develop scientific ideas, and students need to understand that there are many steps along the way to understanding. By revisiting their initial ideas and recognizing how their thinking has changed, students come to realize that they are the ones doing the learning, and that learning cannot be done to them. Formative assessment helps students take ownership in their learning.

## Technology and Formative Assessment

There are many ways teachers can use technology tools to probe for student understanding. Robert Miller, a fifth-grade teacher from Port Orange, Florida, has created a set of outstanding digital videos of some of the *Uncovering Student Ideas in Science* series probes. Students watch the video and then commit to their answer choice using an online polling system. They then discuss the reasons for their answer choice using the EdModo online discussion format. Students explain and argue their ideas online, providing a transcript for the teacher to analyze and use for next steps in designing instruction that will support conceptual understanding. You can view Robert's videos at *http://tinyurl.com/pbp64vk*.

Another technique Robert uses is to take images with a digital camera of natural phenomena around the school campus or in the community that students pass by every day and use the images to elicit students' ideas. For example, before developing the concept of mechanical weathering, a photo might be taken of stones steps at the school that are worn down in one area but not the other. The image is put up on a screen for students to look at and the question is asked, "What do you notice about this photo?" Students soon notice the area that is worn down. The teacher asks, "Why do you think this area is worn down more?" Students share their ideas and then the teacher introduces the concept of mechanical weathering and links it to mechanical weathering of rocks and landforms. The reason Robert uses the photos of familiar places around the school campus or community to elicit and develop students' ideas is that from then on, whenever students pass by that previously unnoticed object or phenomenon, they will see it and think about the concept—a great technique for maintaining conceptual interest and understanding!

**Figure 9.5. Sample Framework With Integrated Assessment**

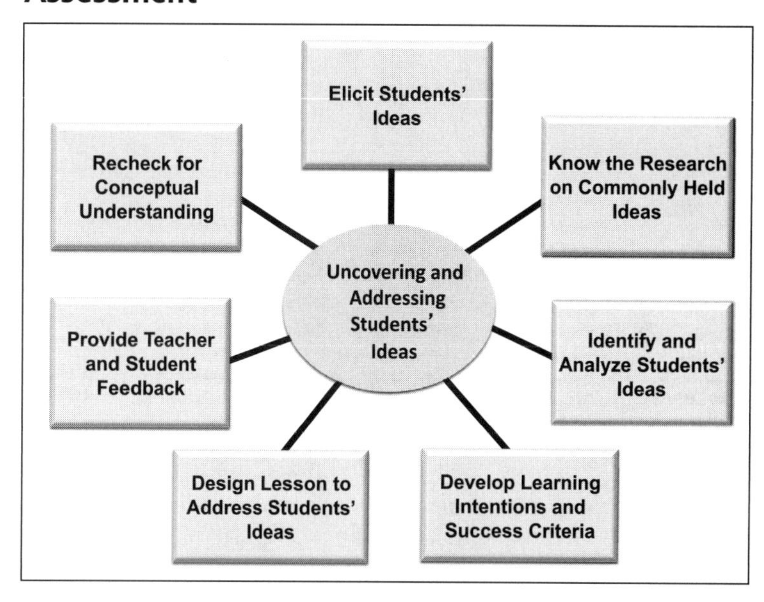


## A Sample Lesson Framework for Uncovering and Addressing Students' Ideas

Assessment is not formative unless the information is used to inform instruction. Figure 9.5 shows a framework that can be used to integrate assessment and instruction into a lesson or set of lessons that target learning goals in which students are likely to have commonly held ideas that have been studied by researchers. We will explore each step.

1. *Elicit Students' Ideas.* Before planning the lesson, uncover the ideas students bring to their learning based on their prior knowledge or experiences. Choose a formative assessment probe, mystery story, or other formative assessment classroom technique (FACT) that will reveal students' ideas related to the standard or performance expectation that the lesson will help prepare students to meet.

   Example: A fourth-grade teacher is designing instruction that will prepare her students to meet the *NGSS* 4-PS4-2 performance expectation: Develop a model to describe that light reflecting from objects and entering the eye allows objects to be seen (NGSS Lead States 2013). Since students need to have a concept of light reflection from ordinary objects in order to prepare them for this performance expectation, she decides to use a formative assessment probe to uncover their ideas about reflection. She selects the probe "Can It Reflect Light?" (Keeley, Eberle, and Farrin 2005).

**Figure 9.6. "Can It Reflect Light?" Probe**

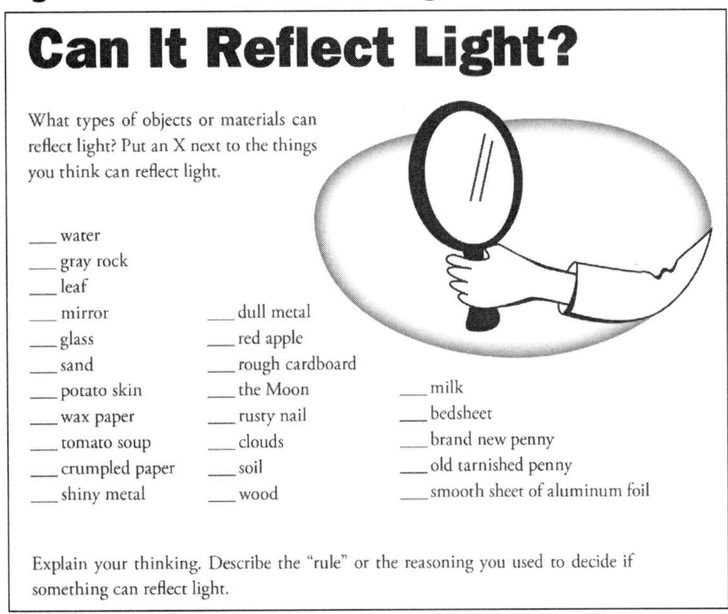

# Can It Reflect Light?

What types of objects or materials can reflect light? Put an X next to the things you think can reflect light.

___ water
___ gray rock
___ leaf
___ mirror
___ glass
___ sand
___ potato skin
___ wax paper
___ tomato soup
___ crumpled paper
___ shiny metal
___ dull metal
___ red apple
___ rough cardboard
___ the Moon
___ rusty nail
___ clouds
___ soil
___ wood
___ milk
___ bedsheet
___ brand new penny
___ old tarnished penny
___ smooth sheet of aluminum foil

Explain your thinking. Describe the "rule" or the reasoning you used to decide if something can reflect light.

*Source:* Keeley, Eberle, and Farrin 2005.

2. *Know the Research on Commonly Held Ideas.* Review the research on commonly held ideas. If you are using a formative assessment probe from the *Uncovering Student Ideas in Science* series or a story from *Everyday Science Mysteries,* research summaries are included in the teacher background material. By becoming familiar with the research, before analyzing students' responses, you will be better prepared to understand students' thinking.

   Example: The teacher reads the Related Research summary provided in the teacher notes for the probe "Can It Reflect Light?" (Figure 9.6). She learns that studies revealed that students know that mirrors reflect light but may not realize that ordinary objects also reflect light. As she reads further, she finds out that research has shown that students fail to connect light reflection to how we see objects. She learns that students have several different models that explain how we see: (1) light goes into the eye and helps us see better; (2) "rays" go from an object into the eye; (3) something goes from the eye to the object; (4) light in the room shines on an object so you can see it; (5) something goes back and forth between and object and the eye; (6) there is no role for light—the eye by itself is an active visual system; (7) light reflects off an object and goes to the eye where the image of the object is formed (the correct scientific idea).

3. *Identify and Analyze Students' Ideas.* Decide how you will use the selected formative assessment classroom technique (FACT), probe, or mystery story to elicit the students' ideas. Collect the formative assessment data and analyze it for evidence of conceptual understanding, partial understanding, or misunderstanding, including commonly held ideas described in the research.

   Example: The teacher decides to use the "Can It Reflect Light?" probe as a card sort. She divides the students up into small groups and gives each group a pack of 24 cards containing the objects listed on the probe. The students are asked to sort the cards into two columns: (1) things that can reflect light and (2) things that cannot reflect light. As they sort each one, they are reminded to discuss their reasons for putting the card into one of the columns. The teacher circulates through the class as the students are discussing their ideas, listening for evidence of research-identified commonly held ideas. As she listens to students, she visually notices that almost all of the groups placed cards in both columns. As she listens to them share their reasoning, she notices that most of the students think an object needs to be shiny or light-colored in order to reflect light. She uses this information to plan the lesson that will confront them with their ideas, and move them toward accepting the idea that ordinary objects can reflect light.

4. *Develop Lesson-Specific Learning Goals.* Develop a lesson-specific learning intention that will focus the lesson or set of lessons and help students know what it is they are expected to learn during the lesson(s). Additionally, develop a set of success criteria that students can use for self-assessment in moving toward an understanding of the learning intention.

   Example: The teacher considers the performance expectation students will eventually need to meet, the initial ideas they have that were revealed during the card sort activity, and identifies two lessons that will help students develop a conceptual understanding of light reflection and how we see objects. She uses this information to develop two learning intentions that will focus instruction on building a bridge between the students' ideas and what they will need to learn in order to meet the performance expectation. She also develops success criteria that the students can use to determine how well they are meeting the learning intention during the lesson.

   Learning Intention #1: Understand how light is reflected from objects.

   - I can show what happens when light strikes a mirror.
   - I can show what happens when light strikes objects that aren't like mirrors.
   - I can draw a picture and use it to explain how light reflects off different kinds of objects.

   Learning Intention #2: Understand the role of light in how we see objects.

   - I can draw a diagram that shows the path of light to and from an object.
   - I can use my model (the diagram) to explain how we see objects.
   - I can use my model to describe what happens when you look at an object in a totally dark room.

5. *Design Lesson to Address Students' Ideas*. Develop a lesson or set of lessons that will help students develop conceptual understanding. Choose instructional strategies that will lead students to discover for themselves the discrepancy between their initial ideas and the scientific understanding.

   Example: After sharing the first learning intention and success criteria with her students, the teacher launches the class into an investigation that will confront them with their initial ideas. She is able to darken her room so students can use flashlights to observe the light reflected from an object onto a shiny whiteboard. Since the students all agreed that the mirror can reflect light, they used that as evidence to show that light from a flashlight can bounce off an object and appear as a spot of light on the whiteboard. The students stood close to the whiteboard and held their mirrors pointed toward the whiteboard. They then pointed a small pen flashlight at the mirror and saw the spot of light reflected from the mirror onto the whiteboard. They repeated this with a piece of aluminum foil and also saw the spot of light reflected onto the whiteboard. The teacher then had students try it with the gray rock. The students were sure that the gray rock could not reflect light because it was not shiny like the other two objects. They held the rock close to the whiteboard, pointed the flashlight at the rock, and sure enough, light could faintly be seen striking the whiteboard. It was more scattered this time—not as concentrated as the mirror. Nevertheless, light did strike the whiteboard, which was evidence that the gray rock could reflect light. The students repeated this with some of the other objects. They changed their initial claims, now recognizing that all of the objects on the cards can reflect light.

6. *Provide Feedback.* Help students recognize the extent to which they met the learning intention and success indicators. Provide opportunities for them to give and receive feedback on their thinking.

   Example: The teacher referred students back to the first learning intention and success criteria. The students turned and talked to a partner, sharing how well they thought they met the success criteria for the lesson and gave feedback on each other's ideas. The teacher then checked with the class on the extent to which they thought they now met the learning intention and determined they were ready for the next lesson and learning intention #2. This is the lesson that would explain how objects are seen by their reflected light. The teacher started the lesson by having students share their ideas about how they see an object. She had them draw pictures to explain how their eyes can see a book sitting on the desk in front of them. The students shared their pictures and explanations. The teacher noted how several of their ideas were similar to the ones described in the research summaries. She then shared the learning intention #2 for the lesson and the success criteria so the students knew what they were expected to learn. She then drew a picture to show the students how we see objects by their reflected light and connected it to the card sort activity and investigation from the previous lesson. She had them talk to a partner and share how their drawings were similar to or different from the teacher's drawing. She explained that their drawings are an example of a conceptual model that represents their thinking. To further help them develop their conceptual understanding, she explained how light from a light source must reflect off the object and enter our eye in order to see it. Since we can see an object, it is evidence that the object reflects light. She then had students revise their drawings so that their models represented the scientific way of explaining how we see an object by its reflected light. Students critiqued each other's models using constructive feedback. At the end of the lesson, the class reviewed the learning intention and success criteria. Some of the students were not sure about the last success criteria. The teacher made a note of that and decided to probe further to find out if students thought they could see things in the dark.

7. *Recheck for Conceptual Understanding*. Use additional FACTs to check for conceptual understanding. Use argument to strengthen newly acquired content knowledge. Revisit students' initial ideas and help students recognize how their thinking has changed. The key to success in developing conceptual understanding is ensuring that students are constructing or reconstructing their new knowledge so that it is compatible with the scientific view. It is important to check for this before moving on to the next set of lessons. This is also a time to check to see if students are ready for summative assessment.

   Example: The teacher decides to use "Apple in the Dark" (Figure 9.7) to probe further to find out if students understand that light must be present in order to see an object and if they can connect it to the models they drew to explain how objects are seen by their reflected light (Keeley, Eberle, and Farrin 2005). Most of the students recognize that that you will not be able to see the apple. Several reproduced their drawings to support their explanation. One student even wrote, "No light, no sight!" The teacher concluded, supported with evidence from formative assessment, that the students were ready for the next lesson and developed the conceptual understanding of light reflection they would need to meet the learning goal.

**Figure 9.7. "Apple in the Dark" Probe**

**Apple in the Dark**

Imagine you are sitting at a table with a red apple in front of you. Your friend closes the door and turns off all the lights. It is totally dark in the room. There are no windows in the room or cracks around the door. No light can enter the room.

Circle the statement you believe best describes how you would see the apple in the dark:

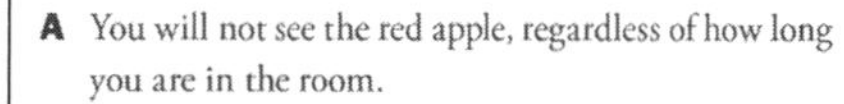

**A** You will not see the red apple, regardless of how long you are in the room.

**B** You will see the red apple after your eyes have had time to adjust to the darkness.

**C** You will see the apple after your eyes have had time to adjust to the darkness, but you will not see the red color.

**D** You will see only the shadow of the apple after your eyes have had time to adjust to the darkness.

**E** You will see only a faint outline of the apple after your eyes have had time to adjust to the darkness.

Describe your thinking. Provide an explanation for your answer.

*Source:* Keeley, Eberle, and Farrin 2005.

Formative assessment includes a variety of ways to link assessment, instruction, and learning so that teachers can make better instructional decisions that support conceptual understanding, while at the same time, promoting learning. A substantive body of research supports formative assessment as one of the best ways to improve student learning. The acid test to determine whether formative assessment is contributing to conceptual understanding in science is not in the number and kinds of techniques teachers use, but in the extent to which they and their students are actively engaged in building a bridge that begins with their initial ideas and involves carefully thought out learning experiences and feedback that will carry them over the bridge to the scientific ideas that help all students make sense of their natural world.

## Questions for Personal Reflection or Group Discussion

1. The so-called high stakes testing has become important to school districts around the country. Do you feel that using formative assessment will improve test scores? Why or why not?
2. Formative assessment is often thought of as a way to find out what students are thinking. What are some other purposes of this type of assessment?
3. Assessment is not formative unless the data is used to inform instruction. Is merely uncovering students' ideas considered formative assessment? Why or why not?
4. Suppose you give your students a probe such as the "Needs of Seeds" probe (Keeley, Eberle, and Tugel 2007), and most of your students think seeds need fertilizer to germinate. How would you proceed to help them to change their idea?
5. Examine the suggestions for getting started with formative assessment. Which of these suggestions do you already incorporate into your practice? Which ones do you think you need to work on?
6. Konicek-Moran's mystery stories and Keeley's probes are somewhat different in their approach to formative assessment. How are they alike and what attributes do they share?
7. Why is it important for the teacher to know the research findings about children's understanding of a concept before using a formative assessment probe or technique?
8. How important is it to have students revisit their previous ideas once they have changed their minds?
9. How does the Lesson Framework for Uncovering and Addressing Students' Ideas compare with your instructional planning? How can you use this framework to plan your lessons?

10. Choose one "golden line" from this chapter (a sentence that really speaks to or resonates with you). Write this on a sentence strip and share it with others. Explain why you chose it.
11. What was the biggest "takeaway" from this chapter for you? What will you do or think about differently as a result?

## Extending Your Learning With NSTA Resources

1. Build a repertoire of effective formative assessment classroom techniques (FACTs), selecting from the combined 125 techniques in *Science Formative Assessment: 75 Strategies for Linking Assessment, Instruction, and Learning* (Keeley 2008) and *Science Formative Assessment: 50 More Strategies for Linking Assessment, Instruction, and Learning* (Keeley 2014).
2. The *Uncovering Student Ideas in Science* series by Page Keeley contains more than 300 formative assessment probes you can use to find out what your students are really thinking about core ideas in science, and you can use that information to make better instructional decisions. In addition to the probes and the teacher notes that accompany each probe, each of the books in this series contains an introductory chapter that provides a wealth of information that can build your formative assessment literacy as well as deepen your understanding of students' ideas in science.
3. The *Everyday Science Mysteries* series by Dick Konicek-Moran can be used to reveal students' ideas as they engage in reading and discussing the story, prior to launching into inquiry. The teacher notes also provide a wealth of background material to help you understand children's ideas and strategies for addressing them. Several of the mystery stories are also linked to a formative assessment probe from the *Uncovering Student Ideas in Science* series.
4. Search the NSTA journal archives for articles on formative assessment. There are many articles in the journals that address ways that formative assessment is used to uncover and understand students' thinking.
5. *What Are They Thinking? Promoting Elementary Learning Through Formative Assessment* (Keeley 2014) contains 30 chapters that describe how teachers use a formative assessment probe, combined with a technique. The vivid classroom vignettes and examples of student work help you see how formative assessment is used in the K–5 classroom. Each of these chapters is based on one of the monthly columns Page Keeley writes for the *Science and Children* journal.
6. *Uncovering Student Ideas in Primary Science: 25 New Formative Assessment Probes*, by Page Keeley, is written specifically for use with PreK–2 children. It uses techniques to engage young children in talking about their ideas in science and provides extensive background material on understanding young children's thinking.

7. Chapters in *Seamless Assessment in Science—A Guide for Elementary and Middle School Teachers* by Sandra Abell and Mark Volkmann provide examples of using formative assessment parallel with inquiry-driven instruction.
8. Check the NSTA Learning Center web seminar archives for Page Keeley's web seminars on formative assessment.
9. Visit *www.uncoveringstudentideas.org* for information on professional development, formative assessment resources, and new book releases.

# Chapter 10

## What Role Does Informal Education Have in Developing Conceptual Understanding?

The role of informal education in science as well as other fields is a fairly recent area for public study. Although opportunities for informal education have been readily available for many years, it is only recently that educational researchers have become interested in finding out how informal education impacts immediate as well as the lifelong learning of children and adults. The National Research Council (NRC) has published two works that give a broad overview of informal science learning opportunities (NRC 2009; Fenichel and Schweingruber 2010). The editors of *Learning Science in Informal Environments: People, Places and Pursuits* have identified six "strands" that participants could, and hopefully would, experience in out-of-school environments:

- Strand 1: Experience excitement, interest, and motivation to learn about phenomena in the natural and physical world.
- Strand 2: Come to generate, understand, remember, and use concepts, explanations, arguments, models, and facts related to science.
- Strand 3: Manipulate, test, explore, predict, question, observe, and make sense of the natural and physical world.
- Strand 4: Reflect on science as a way of knowing; on processes, concepts, and institutions of science; and on their own process of learning about phenomena.
- Strand 5: Participate in scientific activities and learning practices with others, using scientific language and tools.
- Strand 6: Think about themselves as science learners and develop an identity as someone who knows about, uses, and sometimes contributes to science. (NRC 2009, p. 4)

For a person with even a small cognizance of the natural world, it is impossible to walk out of the door at any time of day or night and not be engaged with informal science. If one notices the sky, the Moon, the stars, a sunrise or sunset, the rain, or any of the organisms that surround us in our everyday lives, questions and emotions arise that stimulate the mind, the body, and the soul. Informational and social media, institutions, and the written media all exist to stimulate the mind to delve into an understanding of the world around us. We view education as a "womb to tomb" endeavor. Many of the ideas listed here will be suitable for adults as well as children.

When we think of informal education, often the first things that come to mind are the various museums that exist throughout our world. Dick can speak from personal experience on this topic. If it had not been for informal educational opportunities in the 1930s and 1940s, it is doubtful that he would have had anywhere near a science education that would have led him to his career as a science educator.

## Author Vignette

During the 1930s and 1940s I went to an elementary K–8 school in Chicago, Illinois. Our school, although it was a fine school, had no science program. I was always interested in science, probably because I spent most of my early childhood summers on my grandfather's farm playing around tractors, threshers, and animals and observing nature in a wonderful rural setting. Milking cows, running a cream separator, watching my uncles distill peppermint oil from the peppermint harvest, and smelling the sweet, pungent odor of the peppermint as it wafted over the farm at harvest time were experiences that will never fade from my memory.

One elementary school teacher noticed my interest in things scientific and gave me her science scrapbook consisting of articles she had cut from newspapers and magazines. I would pore over this many times over the years imagining the things it predicted—which now have become reality.

Instead of a science curriculum, our district superintendent decided that our school would pilot a program called Home Mechanics. This program consisted of boxes of things that managed to capture our interest; locks, faucets, irons, tools, and other larger machines such as sewing machines and ovens. I now suspect that since our school was located in a rather poor neighborhood, near the stockyards, it was thought that we would profit more from things that taught us to repair things at home than the "lofty ideas" of science. After all, few, if any of us, were expected to go much further in our education than high school. We learned to sew, iron, cook, replace washers in faucets, and take locks apart—although our house was never quite safe after that due to the many locks that lay around in pieces! No matter how careful I was, there always seemed to be one piece left over when I tried to put the lock back together.

With the support of my parents, I was able to go to the Rosenwald museum in Jackson Park, a holdover from the 1893 Columbian Exhibition. Even though Julius Rosenwald, the CEO of Sears Roebuck and Company and a philanthropist, did not want his name on the museum, we called it that for years, even after it became the Museum of Science and Industry, the largest and first interactive science museum in the western hemisphere. For one dime, I could take the bus to and from the museum each weekend and, armed with a sandwich for lunch, I could spend all day, for free, exploring the working coal mine, wandering through the captured U-505 German submarine, watching chicks hatch from their eggs, studying the transparent human body, and interacting with over 2,000 exhibits (and I do mean *interacting*, for that was the design of the museum). The museum was my science teacher, for I did not have a school science class of any value until I was a sophomore in high school, when Miss Goe taught me to love biology.

Informal science in the right situations can at least motivate a child to want to know more about the world. In this city of museums, my parents made sure I went to each new program at the Adler Planetarium, the Museum of Natural History, and the Art Institute. Yes, parents and informal educational institutions play an important part in the education of both children and adults. We would like to explore some of the aspects of informal education and look at its importance in our national and international lives.

—Dick Konicek-Moran

In this day and age of technology, informal education is everywhere. On radio and television one can watch such programs as *NOVA, Nature, Through the Wormhole,* and all sorts of movies that have scientific themes. Of course, we can search just about any topic on a favorite search engine and find maybe even more than we wanted to know. You may want to visit *whyville.com* to play games or interact with your own avatar in special places like coral reefs or engage in motion activities. However, as teachers you need to be careful which websites you choose: We have to vet them for accuracy and appropriateness. If you are a member of NSTA and receive *NSTA Reports*, you can read Jacob Clark Blickenstaff's articles on using media as discussion starters leading to better understanding of scientific concepts through criticism.

On the radio, Public Radio International (PRI) brings us news and information from all over the world, and such programs as *Science Friday* and *PBS NewsHour* are full of fascinating new findings from the world of science, as are the periodicals such as *Science, Scientific American,* and others brought to us by reliable scientific organizations. And, of course, there is one of the most important and oldest forms of informal education available, the public libraries, formed in this country as early as the 18th century. Interesting programs such as Cornerstones of Science connect libraries with scientists and informal science educators and promote public awareness of science through library programs (*www.cornerstonesofscience.org*).

Then, of course, there are the national and state parks that provide a look into wilderness, and the beauty of natural places. Dick and his wife are still enjoying their tenure (15 years) as volunteers in the park (VIPs) in the Everglades National Park, where they are able to take more adventuresome visitors off the highway into the depths of a cypress dome (they call it the "Cypress Cathedral") where they can wade through the knee-deep water to explore questions that have kept us enthralled over the years. In fact, a biology teacher from Indiana has brought his current classes down each year for the last 11 years to immerse them in the beauty and mystery of one of the last frontiers in North America. Here they volunteer and help protect the natural beauty of the environment and yes, slog into the Cypress Cathedral, to experience firsthand the blazing red bracts of the epiphytes glistening on the trees, while the ibis and herons watch from above. For younger children, the national parks have developed a Junior Ranger Program complete with booklets that guide visiting children and their parents, helping them to look for salient things in each park, but trying not to tell them *what* to see. Children often come back to the visitor center with really good questions as well as to collect their badge.

## Author Vignette

My husband and I had the unique experience of meeting up with Dick and his wife, Kathleen, for a special guided tour slogging through the waist-deep water in the "River of Grass." As we waded through the wetland, I observed mats of scum on the water. It was through Dick's description that I moved from "scum" to understanding that it is what is called periphyton, an aggregate community of algae, cyanobacteria, other microbes, and detritus that is critical to nutrient recycling in the Everglades ecosystem. Being the extraordinary informal educator that he is, Dick began to recite a poem off the top of his head that he had composed about periphyton. Fortunately, I had my camera on me to capture that precious moment that expanded my understanding of this unique Florida ecosystem. It is often informal moments like these that contribute to our understanding of the natural world. You can view the video at *www.youtube.com/watch?v=_VhJ_2LTluc.*

—Page Keeley

Most parks have an environmental education program and work with local schools to make sure that children and teachers benefit from these national treasures. It might be of interest to note that a great many people from the cities surrounding the parks are fearful about entering a wilderness area. This, of course, means that the children in those families inherit the same cautiousness about being in and becoming part of their natural habitat. In fact, children are becoming so "plugged in" in their lives that they are suffering from what author Richard Louv has called "nature-deficit disorder" (Louv 2005). His book *Last Child in the Woods: Saving Our Children From Nature-Deficit Disorder* highlights the dangers that children of our electronic generation face by staying inside because that's where the electrical outlets are. In fact, Louv and several others formed an organization, the Children and Nature Network, to emphasize the importance of being in natural environments.

Science curriculums have also developed with a non-school public in mind; for instance the Outdoor Biology Instructional Strategies (OBIS) curriculum was developed at the Lawrence Hall of Science in Berkeley, California, in the 1970s. The curriculum (still available) has 97 different activities designed for use by community leaders, scouts, parents, and certainly teachers, and are designed to be useful despite any lack of sophistication on the part of adult leaders. As the name implies, the activities are to be used outdoors and provide interesting and fun activities for children of all ages as they explore topics pertaining to biology and the outdoors. Project Learning Tree and Project WET also provide excellent opportunities for teachers and students to investigate and understand the natural world.

There's a little bit of a caveat to put in here: With all of these informal resources at their fingertips, it is no wonder that children come to school with firmly developed theories about their world. We remember the young boy from the *A Private Universe* program who misinterpreted what he learned from a TV presentation on the bat's echolocation as a model for sight and nocturnal animals with "big eyes," thus forming a personal model for how we see as originating in the eyes of the beholder rather than reflection of light from an object into the eye. So, as we have mentioned so many times before—it is *extremely* important to ask students to talk about what they know.

We realize that there a multitude of informal learning opportunities available, so if we neglected to mention your favorite ones, we apologize. One thing we know, we never stop learning and searching for ways to interpret our natural world, regardless of age. The informal science learning opportunities will undoubtedly continue to increase and in this electronic information age, they will certainly become an even more important part of science education.

## Questions for Personal Reflection or Group Discussion

1. How can we help students to view the information available to them on TV or the internet with skepticism and to evaluate their sources for reliability?
2. What was your favorite informal type of informal educational experience as you were growing up? How did it contribute to your understanding of science?
3. What types of informal experiences do you enjoy as an adult learning? How does your experience contribute to your lifelong learning?
4. What is your experience in watching children in a museum? In your opinion how do children "use" the museum's activities to learn about a topic?
5. Dick's experience in the museum over time was quite different than most field trips, which take place in one visit. How do you think you can use a single trip to a museum or National Park as a true and valuable learning experience?
6. What was the biggest "takeaway" from this chapter for you? What will you do or think about differently as a result?

## Extending Your Learning With NSTA Resources

1. NSTA journals often feature articles on informal science learning. Using your NSTA membership number, you can search the archives of all the journals for "informal science."
2. Attend an NSTA area or national conference. There are numerous sessions presented by informal educators, field trips to museums, and an exhibit hall featuring a multitude of informal science organizations.
3. NSTA has a position statement on the importance of informal science. You can read this position statement at *www.nsta.org/about/positions/informal.aspx*.
4. Read about examples of science learning in informal environments in Yager, R., and J. Falk, eds. 2007. *Exemplary science in informal education settings: Standards-based success stories*. Arlington, VA: NSTA Press.
5. NSTA's Science Matters website includes tips for teachers and parents on supporting learning through out-of-school experiences: *www.nsta.org/sciencematters*.

# Appendix

## Putting It All Together: "Balancing" Case Study

To illuminate the three dimensions and the four strands in "science as practice" and emphasize several points made in this book, we offer this case study, which employs new ways of thinking about teaching. In this case study, Deb, a veteran teacher who has faithfully applied ideas from the past on conceptual change to her teaching, decided to apply the new ideas of scientific practices and incorporate the four science learning strands into her curriculum and everyday teaching strategies. Let's see how the interweaving of these strands works in Deb's classroom.

Picture a fourth-grade classroom. The concept under scrutiny is what physicists call the Law of Moments. As a fourth-grade concept, the students were investigating balanced and unbalanced forces. The teacher, Deb, had created a set of balance beams for each group of four children. She had cut half-inch, half-round molding into short pieces and then distributed flat rulers to each group along with a set of identical washers. The children were asked to place the rulers on the half round and make the ruler balance (Figure A.1). The children found that the rulers rested with equal lengths of ruler on each side of the fulcrum. To make the materials more user-friendly, she had also painted a bit of rubber cement onto each half round to make it less slippery.

**Figure A.1. Balance Example**

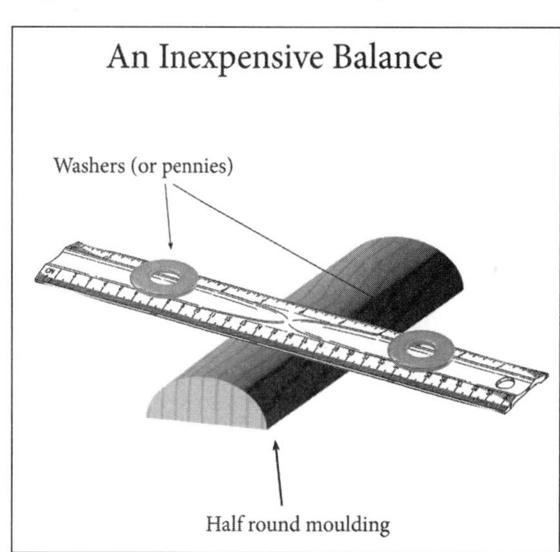


The children played with the miniature "seesaws" for a while, and when they had satisfied themselves that they could balance the ruler, Deb asked them to try putting washers on various spots on the ruler to see if they could come up with some rules about how the ruler might balance on the fulcrum. In the meantime she moved about the room and listened to the conversations and watched the process of exploration at each station. (Practice: Planning and Carrying Out Investigations)

The students began by placing one washer on one side and then placing another on the other side at the same distance from the balance point. Deb then asked the students to talk amongst themselves in their groups and come up with a rule for balancing the ruler. (Practice: Constructing a Scientific Explanation)

She called on Jose, who hardly ever contributed, and he said that his group had come up with the rule, "To balance the ruler on the fulcrum, you place the same number of washers on one side as you do on the other and at the same distance from the balance point."

"Wow, Jose, that is a good rule. I'll write that on the sheet of paper here, which will contain our rules as we discover them."

The sheet was labeled "Our Best Thinking About Balancing, NOW."

"Now class," Deb continued, "I would like you to try putting unequal numbers of washers on each side of the ruler and see if you can come up with a rule that will handle that situation."

The groups went right to work, and since some of them had already done some exploration along these lines, it did not take long before hands shot up in every group. Deb called on the group that had raised their hands first and asked them what they had done.

"We tried putting two washers on one side and one on the other. But the one washer on the opposite side had to be placed further out on the ruler than the two washers. Then we tried it with two washers and three washers and we found that if we put two washers on space one from the middle, we had to put one washer two spaces out. Then we tried three out two spaces, we had to put three washers on space two! So our rule is, 'The more washers inner balance the less washers outer.'" (Practice: Using Mathematical and Computational Thinking)

"That's an interesting rule too. Did anyone else find a similar rule?" Deb wrote this new rule on the paper. Then she asked the group if her writing was accurate for what they had offered as a rule. She did this each time a new rule was offered, since she wanted to make sure that she had not inserted her own thinking into the student offerings.

After a pause another hand went up. "Our group did the same thing and we think that we have a rule that will let you predict exactly where to put the washers when there are different numbers on each side. We tried two out one and one out two. We then tried three out two and two out three. So we continued and put three out four and four out three. They all balanced. So we think the rule should be, 'To balance unequal numbers of washers on the balance you add the number of washers to the number of the space they are on one side and then put the other number of washers on the space so the numbers add up to be equal.'"

"What is your evidence for that rule, Sarah?" (Practice: Analyzing and Interpreting Data)

"Well, two plus one equals one plus two. Three plus two equals two plus three, and so on. Those are our data and they work every time." (Practice: Engaging in Argument From Evidence)

"Fine, Sarah. So for your group I will put up your rule and let the other children find out if it works for them."

The class tried this and was able to balance the beam using Sarah's rule and her data.

"Let's keep trying different combinations, class, and see what you find out. Remember to share with the other members of your group and discuss it." (Practice: Obtaining, Evaluating, and Communicating Information)

The class continued to "mess about" with the balance beams and then from one group, a bunch of hands went up, waving excitedly and Deb asked for their findings.

"Sarah's rule worked fine but then we put three washers out two spaces and balanced it by putting one washer out six spaces and it balanced. That added up to three plus two equals five and six plus one equals six. So the rule doesn't always work. So we put four on three and then two on six and it balanced. Seven doesn't equal twelve!" (Practice: Analyzing and Interpreting Data; and Engaging in Argument From Evidence)

"Your evidence seems right on that one," said Deb. "Let's try some more combinations and see if we can find another rule." (Practice: Obtaining, Evaluating, and Communicating Information; and Asking Questions and Defining Problems)

The class went back to work and finally, a hand went up and that group was recognized.

"Okay, what our group found out was that if we MULTIPLY the number of the space by the number of washers, we always get the board to balance. If you have, say twelve on one side, you have to find a combination that makes twelve on the other." (Practice: Engaging in Argument From Evidence; and Using Mathematics and Computational Thinking)

"Let's see your data, and let's see if it provides us with the evidence we need to make a new rule. (Practice: Analyzing and Interpreting Data)

Rosie brought her data sheet up and copied it onto the whiteboard.

"Let's see if we agree with Rosie's data and can make our balance boards work with her data." (Practice: Planning and Carrying Out Investigations)

Indeed, everyone finally agreed that this worked very well, and all were able to predict where to place washers to keep the balances level.

"Okay, let's put Rosie's rule up here too and see what we have learned."

"Is Sarah's rule correct?"

"Nooooo," came the chorus of response from the class.

"Wait a minute," said Deb. "I may be wrong but I thought it worked fine and you all tried it out and said it worked. What is wrong with the rule?"

"It doesn't work all the time," said Ralph. (Practice: Engaging in Argument From Evidence)

"But it does work when you use her numbers," said Nate.

"What are you thinking about your rule, Sarah?"

"I think my rule works only if you use certain numbers, but Rosie's rule works on all of the numbers, even ours. So, I guess Rosie's is the best rule."

"Well, in science there isn't always a best rule, just a rule that works in more cases."

"Yeah," said Sam. "I think I get it. Both rules are right, but one rule works all of the time and the other works just some of the time, so I guess we'd be better off using the one that works all of the time, the multiply one."

"Would you say more about that Sam?"

"Well, sometimes you get a rule that only works some of the time and then you get a rule that works better but in the end, they are both right—only for different kinds of numbers. Maybe it just takes a little longer to come up with the second or maybe even a third rule." (Practice: Asking Questions and Defining Problems)

"I think you have done a wonderful job today exploring this idea of balancing. If we look at the history of science we find that scientists over the years have had rules that worked and then later discovered that it had to be changed because someone discovered a rule that worked at being a better predictor. This is what you did today. You all did a terrific job of looking at patterns and providing data and defending your claims. Bravo!"

"Now I would like you to talk to people in your group about other examples of how this new rule might apply to other things in your lives." She gave them 10 minutes to discuss this, and then called upon the groups for their responses. She expected a lot of playground examples, and she was not disappointed. (Practice: Obtaining, Evaluating, and Communicating Information)

## Analyzing the Teacher's Actions

So, in this case what did the teacher do to involve the students in the scientific practices? First, she encouraged them to try different combinations of washers and spaces and to look for patterns that would lead to a common rule that could be used to predict how the washers were to be arranged on the balance to achieve equilibrium. Second, she asked the students to create data and provide evidence and support for their rules. In the same vein, she asked the class as a whole to validate each rule for themselves, thus building consensus in the class. Third, she allowed the students to see that rules may have designations other than right or wrong and that rules may be modified in order to make them more inclusive. She also alluded to the nature of the history of science showing that this was the way in which scientific knowledge grows. Remember, Piaget did say that knowledge is a series of transformations that become increasingly adequate (1961). We might add that while the children were seeking rules, they were developing theories based on the data that led to evidence, which in turn fed their theories about balancing. As the children explored the data and evidence and developed the rules, each rule became "increasingly adequate." Put them all together and they spell *inquiry*.

Did Deb use the four *learning strands* in this lesson? Indeed, the rules that she had them create were based on their findings with the balance scales, which would come under the learning strand *generating scientific evidence*, and then having them put each other's rules to the test made them *reflect on scientific knowledge* and then *understand science explanations*. The whole lesson had them *participating productively in science*.

In terms of the core disciplinary idea, her students were developing initial ideas about balanced and unbalanced forces using the crosscutting concepts of Patterns and Cause and Effect. She opened the door by asking the students to think of other examples of balancing in their everyday lives. Exploration of these examples would comprise the meat of the coming science explorations into understanding force and motion. Perhaps she would go by way of the path of simple machines. She would have to ponder that one.

Deb made sure that the children in her class were made aware of the importance of sharing with and arguing with other members of their group. This means that Deb made certain that the point in an argument was not "winning" but arguing with civility from evidence. Dialogue among students is so important that it should be stressed in every subject at all grade levels.

# References

Albe, V. 2008. When scientific knowledge, daily life experience, epistemological and social considerations intersect: Students' argumentation in group discussion on a socio-scientific issue. *Research in Science Education* 38 (1): 67–90.

Alexander, R. J. 2008. *Towards dialogic teaching: Rethinking classroom talk*. 4th ed. York, UK: Dialogos.

Alvermann, D. E., and C. R. Hynd. 1989. Effects of prior knowledge activation modes and text structure on non-science majors comprehension of physics. *Journal of Educational Research* 83: 97–102.

American Association for the Advancement of Science (AAAS). 1960. *Science: A process approach*. New York: Xerox Corporation.

American Association for the Advancement of Science (AAAS). 1988. *Science for all Americans*. New York: Oxford University Press.

American Association for the Advancement of Science (AAAS). 2001. *Atlas of science literacy. Volume 1*. Washington, DC: AAAS.

American Association for the Advancement of Science (AAAS). 2007. *Atlas of science literacy. Volume 2*. Washington, DC: AAAS.

American Association for the Advancement of Science (AAAS). 2009. *Benchmarks for science literacy online. www.project2061.org/publications/bsl/online*

American Institute for Biological Studies (AIBS). 1958. *Biological science curriculum study*. Dubuque, IA: Kendall Hunt.

Andre, T., M. Whigham, A. Hendrikson, and S. Chambers. 1999. Competency beliefs, positive affect and gender stereotypes of elementary students and their parents about science versus other school subjects. *Journal of Research in Science Teaching* 36: 719–748.

Andre, T., and M. Windshitl. 2003. Interest, epistemological belief, and intentional conceptual change. In *Intentional conceptual change*, ed. G. Sinatra, and P. Pintrich, 173–197. Mahwah, NJ: Lawrence Erlbaum Associates.

Annenberg Learner. 1995. *Workshop 3, Physics: Hands-on/minds-on learning*. Annenberg Foundation: *www.learner.org/workshops/privuniv/pup03.html*

Atkin, J. M., and R. Karplus. 1962. Discovery or invention? *The Science Teacher* 29 (5): 45–51.

Ausubel, D. 1963. *The psychology of meaningful verbal learning*. New York: Grune and Stratton.

Beck, T., and E. Leishman. 1996. Will plants drink green water? *Educational Leadership* 54 (4): 56–59.

Beeth, M. E., and P. W. Hewson. 1999. Learning goals in an exemplary science teacher's practice: Cognitive and social factors in teaching for conceptual change. *Science Education* 83 (6): 738–760.

# References


Bell, P., L. Bricker, C. Tzou, T. Lee, and K. Van Horne. 2012. Exploring the science framework: Engaging learners in scientific practices related to obtaining, evaluating, and communicating information. *Science Scope* 36 (3): 17–21.

Bell, P., and M. Linn. 2000. Scientific arguments as learning artifacts: Designing for learning from the web with KIE. *International Journal of Science Education* 22 (8): 797–817.

Birbili, M. 2006. Mapping knowledge: Concept maps in early childhood education. *Early Childhood Research & Practice* 8 (2): 1–11.

Black, P., and C. Harrison. 2002. *Science inside the black box; Assessment for learning in the science classroom.* London: Kings College London Department of Education and Professional Studies.

Bowen, M., and A. Bartley. 2013. *The basics of data literacy: Helping your students (and you!) make sense of data.* Arlington, VA: NSTA Press.

Bransford, J. D., A. L. Brown, and R. R. Cocking, eds. 2000. *How people learn: Brain, mind, experience, and school.* Washington, DC: National Academies Press.

Brice-Heath, S. 1983. *Ways with words: Language, life, and work in communities and classrooms.* New York: McGraw-Hill.

Brodsky, L., A. Falk, and K. Beals. 2013. Helping students evaluate the strength of evidence in scientific arguments. *Science Scope* 36 (9): 22–27.

Brooks, M. 2011. *Free radicals: The secret anarchy of science.* New York: Penguin.

Buttemer, H. 2006. Inquiry on board. *Science and Children* 43 (2): 34–39.

Bybee, R. 2002. *Learning science and the science of learning.* Arlington, VA: NSTA Press.

Bybee, R. 2009. The BSCS instructional model and 21st century skills: A commissioned paper prepared for a workshop on exploring the intersection of science education and the development of 21st century skills. Submitted to the National Academies Board on Science Education.

Bybee, R. 2010. *The teaching of science: 21st century perspectives.* Arlington, VA: NSTA Press.

Bybee, R. 2014. The BSCS 5E instructional model: Personal reflections and contemporary implications. *Science and Children* 51 (9): 3–4.

Bybee, R., E. Buchwald, S. Crissman, D. R. Heil, P. J. Kuerbis, C. Matsumoto, and J. D. McInerey. 1989. *Science and technology education for the elementary years: Frameworks for curriculum and instruction.* Washington, DC: The National Center for Improving Instruction.

Carey, S. 2009. *The origin of concepts,* Oxford, UK: Oxford University Press.

Carnegie Corporation of New York and Institute for Advanced Study (Carnegie-IAS). 2009. *The opportunity equation.* New York: Carnegie Corporation of New York.

Chen, Z., and D. Klahr. 1999. All other things being equal: Acquisition and transfer of the control of variables strategy. *Child Development* 70 (5): 1098–1120.

Clement, J. 2008. Six levels of organization for curriculum design and teaching. In *Model based learning and instruction in science,* ed. J. Clement and M. A. Rea-Ramirez, 255–272. New York: Springer.

Clement, J. 2009. *Creative model construction in scientists and students: The role of imagery, analogy, and mental simulation.* New York: Springer.

Clement, J., D. Brown, C. Camp, J. Kudukey, J. Minstrell, K. Schultz, M. Steinburg, and V. Veneman. 1987. Overcoming students' misconceptions in physics: The role of anchoring intuitions and analogical validity. In *The second international seminar on misconceptions and educational strategies in science and mathematics,* ed. J. Novak, 84–97. Ithaca, NY: Cornell University.

Dewey, J. 1910. Science as subject matter and as method. *Science* 31 (787): 121–127.

diSessa, A. A., and J. Minstrell. 2000. Establishing the norms of scientific argumentation in classrooms. *Science Education* 4: 287–312.

Driver, R., H. Asoko, J. Leach, E. Mortimer, and P. Scott. 1994. Constructing scientific knowledge in the classroom. *Educational Researcher* 23 (7): 5–12.

Driver, R., E. Guesne, and A. Tiberghien. 1985. *Children's ideas in science*. Milton Keynes, England: Open University Press.

Driver, R., J. Leach, R. Millar, and P. Scott. 1996. *Young people's images in science*. New York: Open University Press.

Duckworth, E. R. 1987. *The having of wonderful ideas and other essays on teaching and learning*. New York: Teachers College Press.

Duckworth, E. R. 2001. *Tell me more: Listening to learners explain*. New York: Teachers College Press.

Duckworth, E.R. 2012. *When teachers listen and learners explain*. Video. *www.youtube.com/watch?v=1sfgenKusQk*

Duran, E., L. Duran, J. Haney, and A. Scheuermann. 2011. A learning cycle for all students: Modifying the 5E instructional model to address the needs of all learners. *The Science Teacher* 78 (3): 56–60.

Duschl, R. A., and J. Osborne. 2002. Supporting and promoting argumentation discourse in science education. *Studies in Science Education* 18: 39–72.

Eichinger, D., C. W. Anderson, A. S. Palinscar, and Y. M. David. 1991. An illustration of the roles of content knowledge, scientific argument, and social norms in collaborative problem solving. Paper presented at the annual meeting of the American Educational Research Association, Chicago, IL.

Eisenkraft, A. 2003. Expanding the 5E model. *The Science Teacher* 70 (6): 56–59.

Elementary Science Study (ESS). 1972. Education Development Center, Newton, MA: Education Development Center.

Fenichel, M., and H. A. Schweingruber. 2010. *Surrounded by science: Learning science in informal environments*. Washington, DC: National Academies Press.

Firestein, S. 2012. *Ignorance: How it drives science*. New York: Oxford University Press.

Forbus, K., D. Gentner, J. Everett, and M. Wu. 1997. Towards a computational model of evaluating and using analogical inferences. *Proceedings of the 19th Annual conference of the cognitive science society*, 229–234. London: LEA.

Frayer, D., W. Frederick, and H. Klausmeier. 1969. *A schema for testing the level of concept mastery*. Madison, WI: Wisconsin Center for Education.

Gallas, K. 1995. *Talking their way into science: Hearing children's questions and theories, responding with curricula*. New York: Teachers College Press.

Gelman, R., and J. Lucariello. 2002. Role of learning in cognitive development. In *Stevens' handbook of experimental psychology, vol. 3*, ed. H. Pashler and C. R. Gallistel, 395–443. New York: Wiley.

Gentner, D. 1983. Structure-mapping a theoretical framework for analogy. *Cognitive Science* 7: 155–170.

Gilbert, J., and M. Reiner. 2000. Thought experiments in science education: Potential and current realization. *International Journal of Science education* 22 (3): 265–283.

Glen, N., and S. Dotger. 2009. Elementary teachers' use of language to label and interpret scientific concepts. *Journal of Elementary Science Education* 21 (4): 71–83.

Goodall, J. 1985. Personal conversation.

Gordon, W. J. J. 1961 *Synectics: The development of creative capacity*. New York: Harper and Row.

Grahame, K. 1908. *Wind in the willows*. London: Methuen.

Harvard-Smithsonian Center for Astrophysics (producer). 1997. *Minds of Our Own* DVD. St. Louis, MO: Annenberg Learner.

Hattie, J. 2012. *Visible learning for teachers: Maximizing impact on learning*. Thousand Oaks, CA: Corwin Press.

# References


Hawkins, D. 1965. Messing about in science. *Science and Children* 2 (5): 5–9.

Hawkins, D. 1978. Critical barriers to science learning. *Outlook* 29: 3–25.

Haysom, J., and M. Bowen. 2010. *Predict, observe, explain: Activities enhancing scientific understanding*. Arlington, VA: NSTA Press.

Hein, G. E. 1968. Children's science is another culture. *Technology Review* 71 (2): 61–65.

Hennesey, M. 2003. Metacognitive aspects of students' reflective discourse: Implications for intentional conceptual change teaching and learning. In *Intentional Conceptual Change*, ed. G. Sinatra and P. Pintrich, 103–132. Mahwah, NJ: Lawrence Erlbaum Associates.

Hershberger, K., C. Zembal-Saul, and M. L. Starr. 2006. Evidence helps KLW get a KLEW. *Science and Children* 43 (5): 50–53.

Hewitt, P. G. 1971. *Conceptual physics*. New York: Little, Brown and Co.

Hewitt, P. G. 2006. *Conceptual physics*. Upper Saddle River, NJ: Prentice Hall.

Hewson, P. 1992. Conceptual change in science teaching and teacher education. Paper presented at a meeting on "Research and Curriculum Development in Science Teaching," National Center for Educational Research, Documentation, and Assessment. Madrid, Spain.

Hofer, B., and P. Pintrich. 1997. The development of epistemological theories: Beliefs about knowledge and knowing and their relation to learning. *Review of Educational Research* 67: 88–140.

Karplus, R. 1977. Science teaching and the development of reasoning. *Journal of Research in Science Teaching* 14 (2): 169–175.

Keeley, P. 2008. *Science formative assessment: 75 practical strategies for linking assessment, instruction, and learning*. Thousand Oaks, CA: Corwin Press.

Keeley, P. 2010. *Uncovering student ideas in life science, vol. 1: 25 new formative assessment probes*. Arlington, VA: NSTA Press.

Keeley, P. 2013a. Is it a rock? Continuous formative assessment. *Science and Children* 50 (8): 34–37.

Keeley, P. 2013b. *Uncovering student ideas in primary science: 25 new formative assessment probes*. Arlington, VA: NSTA Press.

Keeley, P. 2014. *What are they thinking? Promoting elementary learning through formative assessment*. Arlington, VA: NSTA Press.

Keeley, P. 2015. *Science formative assessment, volume 2: 50 more practical strategies for linking assessment, instruction, and learning*. Thousand Oaks, CA: Corwin Press.

Keeley, P., F. Eberle, and C. Dorsey. 2008. *Uncovering student ideas in science, vol. 3: Another 25 formative assessment probes*. Arlington, VA: NSTA Press.

Keeley, P., F. Eberle, and L. Farrin. 2005. *Uncovering student ideas in science, vol. 1: 25 formative assessment probes*. Arlington, VA: NSTA Press.

Keeley, P., F. Eberle, and J. Tugel. 2007. *Uncovering student ideas in science, vol. 2: 25 more formative assessment probes*. Arlington, VA: NSTA Press.

Keeley, P., and R. Harrington. 2010. *Uncovering student ideas in physical science, vol. 1: 45 new force and motion formative assessment probes*. Arlington, VA: NSTA Press.

Keeley, P., and R. Harrington. 2014. *Uncovering student ideas in physical science, vol. 2: 39 new electricity and magnetism formative assessment probes*. Arlington, VA: NSTA Press.

Keeley, P., and C. Sneider. 2012. *Uncovering student ideas in astronomy: 45 new formative assessment probes*. Arlington, VA: NSTA Press.

Keeley, P., and J. Tugel. 2009. *Uncovering student ideas in science, vol. 4: 25 new formative assessment probes*. Arlington, VA: NSTA Press.

Kennedy, G. 2003. *Progymnasmata: Greek textbooks of prose composition and rhetoric*. Boston: Brill Academic Publishers.

Kern, C., and K. Crippen. 2008. Mapping for conceptual change. *The Science Teacher* 75 (6): 32–38.

Konicek-Moran, R. 2008. *Everyday science mysteries: Stories for inquiry-based science teaching*. Arlington, VA: NSTA Press.

Konicek-Moran, R. 2009. *More everyday science mysteries: Stories for inquiry-based science teaching*. Arlington, VA: NSTA Press.

Konicek-Moran, R. 2010. *Even more everyday science mysteries: Stories for inquiry-based science teaching*. Arlington, VA: NSTA Press.

Konicek-Moran, R. 2011. *Yet more everyday science mysteries: Stories for inquiry-based science teaching*. Arlington, VA: NSTA Press.

Konicek-Moran, R. 2013a. *Everyday Earth and space and science mysteries: Stories for inquiry-based science teaching*. Arlington, VA: NSTA Press.

Konicek-Moran, R. 2013b. *Everyday life science mysteries: Stories for inquiry-based science teaching*. Arlington, VA: NSTA Press.

Konicek-Moran, R. 2013c. *Everyday physical science mysteries: Stories for inquiry-based science teaching*. Arlington, VA: NSTA Press.

Kuhn, D. 1995. Scientific thinking and knowledge acquisition. In *Strategies of knowledge acquisition*, ed. D. Kuhn, M. Garcia-Mila, A. Zohar, and C. Andersen, 152–157. Hoboken, NJ: Wiley.

Kuhn, D., and E. Phelps. 1982. The development of problem-solving strategies. In *Advances in child development and behavior*, ed. H. Reese, 1–44. New York: Academic Press.

Kuhn, M., and M. McDermott. 2013. Negotiating the way to inquiry. *Science and Children* 50 (9): 52–57.

Kuhn, T. 1996. *The structure of scientific revolutions*. 2nd ed. Chicago: University of Chicago Press.

Leach, J., R. Konicek, and B. Shapiro. 1992. The ideas used by British and North American school children to interpret the phenomenon of decay: A cross-cultural study. Paper presented at the Annual Meeting of the American Educational Research Association, San Francisco, CA.

Leahy, S., C. Lyon, M. Thompson, and D. Wiliam. 2005. Classroom assessment: Minute-by-minute and day-by-day. *Educational Leadership* 63 (3): 18–24.

Lemke, J. L. 1990. *Talking science language, learning, and values*. Norwood, NJ: Ablex.

Leroi, A. M. 2014. *The lagoon: How Aristotle invented science*. New York: Viking.

Lindquist, M. M., and V. L. Kouba. 1989. Geometry. In *Results from the fourth mathematics assessment of the national assessment of educational progress*, ed. M. M. Lindquist, 44–54. Reston, VA: National Council of Teachers of Mathematics.

Livio, M. 2013. *Brilliant blunders: From Darwin to Einstein*. New York: Simon & Schuster.

Louv, R. 2005. *Last child in the woods: Saving our children from nature-deficit disorder.* Chapel Hill, NC: Algonquin Books.

Mager, R. 1968. *Developing attitude toward learning*. Palo Alto, CA: Fearon Publishers.

Mascazine, J. R., and W. S. McCann. Conceptual change in the classroom. *Science Scope* 22 (3): 23–25.

McDermott, M., and B. Hand. 2010. A secondary reanalysis of student perceptions of non-traditional writing tasks over a ten year period. *Journal of Research in Science Teaching* 47 (5): 501–517.

McNeill, K. L., and J. Krajcik. 2008. Assessing middle school students' content knowledge and reasoning through written scientific explanations. In *Assessing science learning: Perspectives from research and practice,* ed. J. Coffey, R. Douglas, and C. Stearns, 101–116. Arlington, VA: NSTA Press.

Mercer, N., and K. Littleton. 2007. *Dialogue and the development of children's thinking.* London: Routledge.

Metz, K. E. 1995. Reassessment of developmental constraints on children's science instruction. *Review of Educational Research* 65: 93–127.

Metz, K. E. 1997. On the complex relation between cognitive developmental research and children's science curricula. *Review of Educational Research* 67: 151–163.

Metz, K. E. 2004. Children's understanding of scientific inquiry: Their conceptualization of uncertainty in investigations of their own design. *Cognition and Instruction* 22 (2): 219–290.

Michaels, S., and R. Sohmer. 2001. Discourses that promote new academic identities. In *Discourses in search of members*, ed. D. Li, 171–219. New York: University Press of America.

Michaels, S., A. Shouse, and H. Schweingruber. 2008. *Ready, set, SCIENCE! Putting research to work in K–8 science classrooms.* Washington, DC: National Academies Press.

Millar, R., R. Driver, J. Leach, and P. Scott. 1993. *Students' understanding of the nature of science*. Leeds, UK: Children's Learning in Science Research Group.

Minstrell, J. 1989. Teaching science for understanding. In *Toward the thinking curriculum: Current cognitive research*, ed. L. B. Resnick and L. E. Klopfer, 129–149. Alexendria, VA: Association for Supervision and Curriculum Development.

Minstrell, J. 1992. Facets of students' knowledge and relevant instruction. In *Proceedings of the international workshop on research in physics education: Theoretical issues and empirical studies*, ed. R. Duit, F. Goldberg, and H. Niedderer, 110–128. Kiel, Germany: IPN.

National Governors Association Center for Best Practices and Council of Chief State School Officers (NGAC and CCSSO). 2010. *Common core state standards for English language arts and literacy*. Washington, DC: National Governors Association for Best Practices, Council of Chief State Schools.

National Research Council (NRC). 1996. *National science education standards*. Washington, DC: National Academies Press.

National Research Council (NRC). 2005. *America's lab report: Investigations in high school science.* Washington, DC: National Academies Press.

National Research Council (NRC). 2007. *Taking science to school.* Washington, DC: National Academies Press.

National Research Council (NRC). 2009. *Learning science in informal environments: People, places, and pursuits.* Washington, DC: National Academies Press.

National Research Council (NRC). 2012. *A framework for K–12 science education: Practices, crosscutting concepts and core ideas.* Washington, DC: National Academies Press.

National Science Teachers Association (NSTA) Board of Directors. 2000. NSTA position statement: The nature of science. *www.nsta.org/about/positions/natureofscience.aspx*

NGSS Lead States. 2013. *Next Generation Science Standards: For states, by states.* Washington, DC: National Academies Press. *www.nextgenscience.org/next-generation-science-standards.*

Norton-Meier, L., B. Hand, L. Hockenberry, and K. Wise. 2008. *Questions, claims and evidence*. Portsmouth, NH: Heinemann.

Novak, J. D. 1998. *Learning, creating, and using knowledge: Concept maps as facilitative tools in schools and corporations.* Mahwah, NJ: Erlbaum.

Novak, J., and A. Cañas. 2006. The theory underlying concept maps and how to construct them (Technical Report No. IHMC CmapTools 2006-01). Pensacola, FL: Institute for Human and Machine Cognition.

Nussbaum, J., and N. Novick. 1982. Alternative frameworks, conceptual conflict, and accommodation: Toward a principled teaching strategy. *Instructional Science* 11 (3): 183–200.

Ogle, D. M. 1986. K-W-L: A teaching model that develops active reading of expository text. *The Reading Teacher* 39: 564–570.

Osborne, J. E., S. Erduran, S. Simon, and M. Monk. 2001. Enhancing the quality of argument in school science. *School Science Review* 83 (301): 63–70.

Osborne, R., and P. Freyberg. 1985. *Learning in science: The implications of children's science.* Portsmouth, NH: Heinemann.

Peck, A., and R. D. Konicek. 1998. *Connect* 12 (2): November / December.

Perkins, D. 1993. Teaching for understanding. *American Educator* 17 (3): 28–35.

Perkins, D., and T. Grotzer. 2005. Dimensions of causal understanding: The role of complex causal models. *Studies in Science Education* 41 (1 / 2): 117–166.

Physical Science Study Committee (PSSC). 1960. *Physics*. Lexington, MA: D. C. Heath.

Piaget, J. 1968. *Genetic epistemology*, trans. Eleanor Duckworth. New York: Columbia University Press.

Piaget, J. 1973. *To understand is to invent.* Chicago: Grossman.

Pines, A. L. 1985. Toward a taxonomy of conceptual relations and other implications for the evaluation of cognitive structures. In *Cognitive structures and conceptual change*, ed. L. H. T. West, and A. L. Pines, 101–116. Orlando, FL: Academic Press.

Posner, G. J., K. A. Strike, P. W. Hewson, and W. A. Gertzog. 1982. Accommodation of a scientific conception: Toward a theory of conceptual change. *Science Education* 66 (2): 211–227.

President and Fellows of Harvard University for Understandings of Consequence Project of Project Zero. 2005. *Causal patterns in density: Lessons to infuse into density units to enable deeper understanding.* Cambridge, MA: Harvard Graduate School of Education.

Preszler, R. 2004. Cooperative concept mapping. *Journal of College Science Teaching* 33 (6): 30–35.

Prince, G. 1970. *The practice of creativity*. New York: Harper and Row.

Rea-Ramirez, M. A. 2008. Determining target models and effective learning pathway for developing understanding of biological topics. In *Model based learning and instruction in science*, ed. J. Clement, and M. A. Rea-Ramirez, 45–58. New York: Springer.

Rea-Ramirez, M. A., and M. C. Nunez-Oviedo. 2008. Role of discrepant questioning leading to mode element modification. In *Model based learning and instruction in science*, ed. J. Clement and W. A. Rea-Ramirez, 195–213. New York: Springer.

Reiner, M. 1998. Thought experiments and collaborative learning in physics. *International Journal of Science Education* (20): 1043–1058.

Richhart, R., M. Church, and K. Morrison. 2011. *Making thinking visible: How to promote engagement, understanding, and independence for all learners.* San Francisco, CA: Jossey-Bass.

Rogoff, B., E. Matusov, and C. White. 1996. Models of teaching and learning: Participation in a community of learners. In *Handbook of education and human development*, ed. D. Olson, and N. Terrance, 388–414. Oxford, UK: Blackwell.

Roseman, J. E., S. Kesidou, L. Stern, and A. Caldwell. 1999. Heavy books light on learning: AAAS Project 2061 evaluates middle grades science textbooks. *Science Books & Films* 35 (6): 243–247.

Roth, K. 1990. Developing meaningful conceptual understanding in science. In *Dimensions of thinking and cognitive Instruction*, ed. B. F. Jones and L. Idol, 139–176. Hillsdale, NJ: Erlbaum.

Rowe, M. B. 1974. Wait time and rewards as instructional variables, their influence on language, logic, and fate control. *Journal of Research on Science Teaching* 11: 81–94.

Rowe, M. B. 1986. Wait time: Slowing down may be a way of speeding up. *Journal of Teacher Education* 37: 43.

Ruffman, T., J. Perner, D. R. Olson, and M. Doherty. 1993. Reflecting on scientific thinking: Children's understanding of the hypothesis-evidence relation. *Child Development* 64: 1617–1636.

Samples, R. 1974. Personal conversation.

Sampson, V., P. Carafano, P. Enderle, S. Fannin, J. Grooms, S. A. Southerland, C. Stallworth, and K. Williams. 2015. *Argument-driven inquiry in chemistry: Lab investigations for grades 9–12.* Arlington, VA: NSTA Press.

Sampson, V., P. Enderle, L. Gleim, J. Grooms, M. Hester, S. Southerland, and K. Wilson. 2014. *Argument-driven inquiry in biology: Lab investigations for grades 9–12.* Arlington, VA: NSTA Press.

Sampson, V., and S. Schleigh. 2013. *Scientific argumentation in biology: 30 classroom activities.* Arlington, VA: NSTA Press.

Schauble, L., R. Glaser, R. Duschl, S. Schulze, and J. John, 1995. Students' understanding of the objectives and procedures of experimentation in the science classroom. *Journal of the Learning Sciences* 4 (2): 131–166.

Schauble, L., L. E. Klopfer, and K. Raghavan. 1991. Students' transition from an engineering model to a science model of experimentation. *Journal of Research in Science Teaching* 28 (9): 859–882.

Schneps, M. H., and P. M. Sadler. 1987. *A Private Universe.* Video. Washington, DC: Annenberg Learner.

Schraw, G., K. J. Crippen, and K. D. Hartley. 2006. Promoting self-regulation in science education: Metacognition as part of a broader perspective on learning. *Research in Science Education* 36 (1–2): 111–139.

Schwartz, R. 2007. What's in a word? How word choice can develop (mis) conceptions about the nature of science. *Science Scope* 31 (2): 42–47.

Schwartz, M., P. Sadler, G. Sonnert, and R. Tai. 2009. Depth vs breadth: How content coverage in high school science courses relates to later success in college coursework. *Science Education* 93 (5): 798–826.

Science Curriculum Improvement Study (SCIS). 1970. *Subsystems and variables teacher's guide.* Chicago: Rand McNally.

Scott, P. 1994. Personal conversation.

Shapiro, B. 1994. *What children bring to light.* New York: Teacher's College Press.

Shymansky, J. A., L. V. Hedges, and G. Woodworth. 1990. A reassessment of the effects of inquiry-based science curricula of the 60s on student performance. *Journal of Research in Science Teaching* 27: 127–144.

Sinatra, G. M., and P. R. Pintrich. 2003. The role of intentions in conceptual change learning. In *Intentional conceptual change,* ed. G. M. Sinatra, and P. R. Pintrich, 1–18. Mahwah, NJ: Erlbaum.

Smith III, J. P., A. A. diSessa, and J. Roschelle. 1993. Misconceptions reconceived: A constructivist analysis of knowledge in transition. *The Journal of the Learning Sciences* 3 (2): 115–163.

Sneider, C., K. Kurlich, S. Pulow, and A. Friedman. 1984. Learning to control variables with model rockets: A neo-Piagetian study of learning in field settings. *Science Education* 68 (4): 463–484.

Stepans, J., B. Saigo, and C. Ebert. 1999. *Changing the classroom from within: Partnership, collegiality, and constructivism.* Montgomery, AL: Saiwood.

Sunal, D. W. 1992. The learning cycle: A comparison of models of strategies for conceptual reconstruction: A review of the literature. *http://astlc.ua.edu/ScienceInElem&MiddleSchool/565LearningCycle-ComparingModels.htm*

Thagard, P., and J. Zhu. 2003, Acupuncture, incommensurability and conceptual change. In *Intentional conceptual change,* ed. G. Sinatra and P. Pintrich, 79–102. Mahwah, NJ: Erlbaum.

Vanides, J., Y. Yin, M. Tomita, and M. Araceli Ruiz-Primo. 2005. Using concept maps in the science classroom. *Science Scope* 28 (8): 27–31.

Vasniadou, S. 2003. Exploring the relationships between conceptual change and intentional learning. In *Intentional conceptual change*, ed. G. Sinatra, and P. Pintrich, 407–427. Mahwah, NJ: Erlbaum.

Venville, G., and V. Dawson. 2010. The impact of a classroom intervention on grade 10 students' argumentation skills, informal reasoning, and conceptual understanding of science. *Journal of Research in Science Teaching* 47 (8): 952–977.

Von Glasersfeld, E. 1989. Constructivism in education. In *The international encyclopedia of education supplement, vol.1*, ed. R. Husen and N. Postlethwaite, 162–163. New York: Pergamon Press.

Vygotsky, L. 1962. *Thought and language.* Cambridge, MA: MIT Press.

Vygotsky, L., and A. Luria. 1994. Tool and symbol in child development. In *The Vygotsky reader*, ed. R. van de Veer and J. Valsiner, 99–174. Oxford: Blackwell.

Ward, H., J. Roden, C. Hewlett, and J. Foreman. 2005. *Teaching science in the primary classroom.* London: Paul Chapman Publishing.

Watson, B., and R. Konicek. 1991. Teaching for conceptual change: Confronting children's experience. *Phi Beta Kappan* May (680–684).

Wellington, J., and J. Osborne. 2001. *Language and literacy in science education.* Buckingham, UK: Open University Press.

Wiliam, D. 2011. *Embedded formative assessment.* Bloomington, IN: Solution Tree Press.

Wiser, M. 1988. The differentiation of heat and temperature: History of science and novice-expert shift. In *Ontogeny, phylogeny, and historical development*, ed. S. Strauss, 28–48. Norwood, NJ: Ablex.

Wiser, M., and S. Carey. 1983. When heat and temperature were one. In *Mental models*, ed. D. Genter, and A. Stevens, 267–297. Hillsdale, NJ: Erlbaum.

Wolfe, S. 2006. Teaching and learning through dialogue in primary classrooms in England. Unpublished PhD thesis, University of Cambridge.

Wolfe, S., and R. Alexander. 2008. Argumentation and dialogic teaching: Alternative pedagogies for a changing world. *www.beyondcurrenthorizons.org.uk/wp-content/uploads/ch3_final_wolfealexander_argumentationalternativepedagogies_20081218.pdf*

Yager, R., ed. 2009. *Inquiry: The key to exemplary science.* Arlington, VA: NSTA Press.

# Index

*Page numbers printed in* **boldface** *type refer to figures or tables.*

# Index

# Index

# Index